D1451106

A TALE OF THREE CITIES

(*Medical History*, Supplement No. 9)

William Sharpey

A TALE OF THREE CITIES

THE CORRESPONDENCE OF WILLIAM SHARPEY AND ALLEN THOMSON

edited by

L. S. JACYNA

(*Medical History*, Supplement No. 9)

LONDON
WELLCOME INSTITUTE FOR THE HISTORY OF MEDICINE
1989

ISBN 0 85484 090 7
ISSN 0025 7273 9

Supplements to *Medical History* may be obtained by post from Professional and Scientific Publications, BMA House, Tavistock Square, London WC1H 9JR.

Contents

Illustrations

Acknowledgements

I would like to thank the Keeper of Special Collections of Glasgow University Library for permission to publish the Sharpey-Thomson correspondence; and I am indebted to the staff of the Special Collections Department for much assistance in the preparation of this edition. I am also grateful to Mike Barfoot and Chris Lawrence for their valuable comments and suggestions.

Introduction

Glasgow University Library possesses an extensive collection of the papers of the Thomsons, a family of eminent nineteenth-century medical men. These form part of a larger set of Cullen-Thomson material. Some of these manuscripts were donated to the Library by John Millar Thomson in 1920; the remainder was discovered in the University Library Store in 1973. Among this latter set are a series of letters between Allen Thomson (1809–84) and William Sharpey (1802–80).[1] The bulk of this correspondence consists of letters from Sharpey to Thomson; but there are also several copies and drafts of the letters Thomson sent. These are valuable because Thomson's side of the correspondence has not otherwise survived. Letters from other individuals mentioned in the correspondence are also present. It seems that Thomson brought together all these letters, along with other documents, as sources for the obituary of Sharpey he was writing. A calendar of these letters will be found in the Appendix.

This edition consists only of a selection of the letters. I have chosen those of most interest to the historian of medicine, and, in particular, those that bear on the roles played by Thomson and Sharpey as influential medical scientists and medical politicians in nineteenth-century Britain.

Sharpey is the better-known of the two; he has been the subject of a lengthy article by D. W. Taylor.[2] However, Taylor did not have access to a manuscript biography of Sharpey that Thomson composed in 1880, which provides much new information especially about Sharpey's early career.[3]

He was born at Arbroath in Forfarshire, Scotland in 1802, the stepson of a local medical practitioner. After studying at the public school in Arbroath, Sharpey proceeded in November 1817 to the University of Edinburgh where he first attended the Arts and Natural Philosophy Classes. In the following year he began the medical curriculum; his course of study in the University is outlined below.

1818–19:   Anatomy, Botany
1819–20:   Practice of Medicine, Clinical Medicine; Chemistry
1820–21:   Institutes of Medicine; Materia Medica; Midwifery
1823:   MD *'De ventriculi carcinomate'*
(*Source*: Edinburgh University Matriculation Records)

Thomson noted that he was already in the second year of his studies attending James Gregory's lectures on the Practice of Medicine, and suggested that this indicated that

---

[1] The call-mark of the Thomson MSS is Gen. 1476; most of the Sharpey-Thomson correspondence is in Box 15.

[2] D. W. Taylor, 'The life and teaching of William Sharpey (1802–1880) "Father of Modern Physiology" in Britain', *Med. Hist.*, 1971, **15**: 126–53, 241–59. The other major published biographical source is Allen Thomson's obituary of Sharpey: *Proc. R. Soc. Lond.*, 1880, **31**: xi–xix.

[3] Allen Thomson, [Life of William Sharpey], Thomson MSS, op. cit., note 1 above, Box 16.

Sharpey had "received a considerable part of the rudiments of medical knowledge from his stepfather Dr Arrott".[4]

In addition to his University education, Sharpey attended a number of extra-mural courses, including those of John Barclay in anatomy and John Murray in chemistry. The correspondence reveals that Sharpey also attended lectures by John Thomson (letter 11): according to the list of Sharpey's class tickets Thomson compiled, Sharpey twice attended John Thomson's lectures on military surgery as well as those he gave on diseases of the eye, and the lectures Thomson gave at the College of Surgeons of Edinburgh until 1821.[5] In a testimonial John Thomson wrote for Sharpey in 1836, he stated that "I have been intimately acquainted with Dr Sharpey since 1818 when he first commenced his medical studies."[6]

In the spring of 1821 Sharpey obtained the Diploma of the Royal College of Surgeons of Edinburgh. He then proceeded to London to spend three months studying anatomy at Brooke's School in Windmill Street. Towards the end of 1821 Sharpey went to Paris where he "was closely engaged in the study of Clin. Med. & Surg. at the hospitals".[7]

He then returned to Arbroath where he seems to have been engaged to some degree in the family practice. In answer to a query on this point from Thomson, Sharpey's stepbrother declared in 1880 that "Sharpey was never in medical practice in Arbroath—He may occasionally have taken an interest in any particular patient who came to consult my father but that was all."[8] Thomson appears, however, to have discounted this testimony; there is a hint that James Arrott had reasons of his own to wish to minimize Sharpey's role in the Arbroath practice. Arrott did reveal that "[i]n the beginning of the year 1826 Sharpey had occasion to consider carefully the question of engaging in private practice and resolved not to do so."[9] This "occasion" arose when a local practitioner offered to sell Sharpey his practice. Sharpey declined, according to his nephew, because of a reluctance to enter into competition with "the progeny of old Dr Arrott".[10] Another reason for this decision was, in Thomson's view, a wish "to devote himself to Anatomical and physiological pursuits".[11]

With this end in mind, Sharpey travelled to the Continent in 1827. He initially went to Italy, spending some time in the study of anatomy with Bartolomeo Panizza (1785–1867) in Pavia. In the autumn of 1828 he proceeded to Berlin where he worked under the tutelage of Karl Asmund Rudolphi (1771–1832).[12] Sharpey's own account shows that he spent his time in Berlin engaged in intensive dissection, which he regarded as an essential preparation for the training of an anatomist.

---

[4] Ibid., p. 2.

[5] 'Dr Sharpeys Edin. Univ. Tickets', Thomson MSS, op. cit., note 1 above, Box 16.

[6] A transcription of Sharpey's testimonials is to be found in the Sharpey-Schäfer Collection, Wellcome Institute for the History of Medicine, B. 1–4, ESS/B. 1/5–7.

[7] Thomson, [Life], op. cit., note 3 above, p. 3.

[8] James Arrott to Allen Thomson, 11 May 1880, Thomson MSS, op. cit., note 1 above, Box 16.

[9] James Arrott to Allen Thomson, 11 June 1880, ibid.

[10] William Henry Colvill to Allen Thomson, 3 October 1880, ibid.

[11] Thomson, [Life], op. cit., note 3 above, p. 4.

[12] Ibid., pp. 4–5.

Upon his return to Edinburgh he embarked upon microscopical research. Even as a student Sharpey had shown an interest in minute anatomy, making use of the microscope of the Royal Medical Society.[13] This instrument was, however, uncorrected for chromatic or spherical aberration. Sharpey was one of the first British workers to employ the new achromatic microscopes that became available around 1830.

At the same time he continued his preparations to begin teaching anatomy in the Edinburgh extra-mural school. He spent the summer of 1831 in Berlin collecting specimens and making preparations for this purpose.[14] When he offered his first course in the 1831–2 session it was in partnership with Allen Thomson. As mentioned above, Sharpey had been acquainted with Thomson's father since 1818, when Allen was nine years old. Sharpey was also a student contemporary of John Thomson's other son William, with whom he attended some University classes. It is therefore likely that Sharpey and Allen knew each other from an early date. Thomson had spent the year 1828–9 in Paris. He reported that when he returned to Edinburgh in July 1829, he "found Dr. S. in lodgings in Castle S^r and we then became very intimate and constantly together—in observations &c." It must have been during this period that they decided to enter into a partnership, and Thomson accompanied Sharpey to Germany in 1831 to help in the collection of preparations.[15]

Such partnerships were by no means unusual in the early nineteenth-century Edinburgh extra-mural school: it was common for teachers to join together to offer complementary courses in, for example, surgery and anatomy. Sharpey and Thomson's division of labour was, however, somewhat novel; the former taught anatomy and the latter physiology. John Allen (1771–1843)—after whom Allen Thomson was named—had in 1794 delivered a course of lectures on physiology in Edinburgh;[16] and in 1813 John Gordon—another intimate of the Thomson family—divided his teaching into separate courses of anatomy and physiology.[17] These precedents were not, however, influential; and Sharpey and Thomson's school was unique among contemporary extra-mural teaching establishments. Their lectures were held at 9 Surgeons Square where John and William Thomson, Gordon, and others had also taught.[18] This association persisted for the next four years.

In the summer of 1836 Sharpey became Professor of Anatomy and Physiology at what was then called the University of London, and later became University College. It was at this point that the correspondence began. Sharpey remained in this post until 1874. The correspondence reveals, however, that from the first he had in mind the possibility of returning to Edinburgh as Professor of Anatomy when the opportunity arose (letters 1 and 20). This very nearly came to pass in 1846 (letters

[13] Ibid., p. 3.

[14] Ibid., p. 7.

[15] Allen Thomson, [Chronology of William Sharpey's life], Thomson MSS, op. cit., note 3 above, Box 16.

[16] See [William Thomson], 'Biographical notice', in John Thomson, *An account of the life, lectures, and writings of William Cullen, M.D.*, 2 vols., Edinburgh, William Blackwood, 1859, vol. 1, p. 13.

[17] These innovations are described in C. Lawrence, 'Alexander Monro *Primus* and the Edinburgh manner of anatomy', *Bull. Hist. Med.*, 1988, **62**: 193–214, on p. 213.

[18] Thomson, [Life], op. cit., note 3 above, p. 7.

25–26). Five years later Sharpey again actively considered returning to Edinburgh to fill the Chair of the Institutes of Medicine (letter 47).

During his time in London Sharpey was active in a number of scientific and educational bodies. He was a Secretary of the Royal Society from 1853 to 1872; in this capacity he played an important role in the refereeing of papers submitted for publication in the Royal Society's *Proceedings*.[19] He was also an early appointee to the General Council of Medical Education and Registration established by the Medical Act of 1858. Allen Thomson joined this body in 1859, and several letters are concerned with GMC business (letters 74, 82, 85, 87).

Sharpey's health began to fail in 1871, and in the succeeding years he gradually resigned his teaching and other responsibilities. His final years were blighted by failing eyesight, which was only partially relieved by surgery. He died on 11 April 1880 from bronchitis. Thomson was among those who attended him in his last hours.

All Allen Thomson's biographers agree upon the importance of the circumstances of his upbringing to his future career.[20] He was the younger son of John Thomson, one of the most dynamic figures in the medical life of late eighteenth- and early nineteenth-century Edinburgh: he was Professor of Surgery at the Royal College of Surgeons of Edinburgh, as well as successively holding the Regius Chairs of Military Surgery and Pathology in the University.[21] Allen's elder brother William was also a notable medical teacher.

Thomson had a conventional education, proceeding to the University after study at the Edinburgh High School. The courses he took are listed below.

1824–5:   Chemistry, Humanities II
1825–6:   Anatomy, Chemistry
1826–7:   Chemistry, Botany
1827–8:   Materia Medica, Institutes of Medicine
1829–30:   Practice of Medicine, Midwifery, Clinical Medicine
1830:   M.D. '*De evolutione cordis in animalibus vertebratis*'
(*Source*:   Edinburgh University Matriculation Records)

Like Sharpey, Thomson supplemented these courses by attendance at the extra-mural school and Continental travel. The notes he took of his medical studies in Paris in 1828 and 1829 are preserved among the Thomson papers.[22] As a student in Edinburgh Thomson enjoyed the distinction of being President of the Royal Medical Society.

It was, according to Aitken, John Thomson's "great desire that [Allen] should become a teacher of anatomy, and devote himself to anatomical and physiological

---

[19] See Taylor, op. cit., note 2 above, pp. 241–3.
[20] The chief source on Allen Thomson's life is the obituary by W. Aitken in *Proc. R. Soc. Lond.*, 1887, **42**: xi–xxviii.
[21] On John Thomson see: [William Thomson], op. cit., note 16 above.
[22] I have discussed some of these notes in '*Au lit des malades*: A. F. Chomel's clinic at the Charité, 1828–9', *Med. Hist.*, 1989, **33**: 420–49.

pursuits".[23] From an early stage in his career Thomson himself favoured theoretical rather than practical medicine. He was particularly drawn to embryology, the subject of his MD dissertation. He was—again like Sharpey—a pioneer microscopist and among the first to teach the use of the microscope to students in Edinburgh.[24]

In 1836 Thomson ended his partnership with Sharpey and temporarily gave up teaching because of ill health. He moved to London to become private physician to John Russell, sixth Duke of Bedford; it was from this situation that he sent the first letter in the correspondence. Thomson toured the Continent with the Bedford family and lived with them in the Scottish Highlands and in Ireland for a number of years; however, it is clear from the correspondence that his intention was ever to resume his teaching career in Edinburgh. This he did in the 1837–8 session, this time offering a course of lectures in anatomy.

In 1839 Thomson was appointed Professor of Anatomy at the Marischal College, Aberdeen; but the appointment was short-lived. In 1841 he returned to Edinburgh as an extra-mural lecturer in anatomy. Aitken suggested that Thomson resigned his post in Aberdeen and returned to Edinburgh in the anticipation that the Institutes Chair in Edinburgh was soon to fall vacant.[25] There is nothing in the letters between Sharpey and Thomson of this period to confirm this conjecture. The letters do, however, suggest another motive for Thomson's departure from Aberdeen: his classes there were disappointingly small (letter 17). Nevertheless, Thomson was appointed Professor of the Institutes in 1842 following William Pulteney Alison's resignation. But it is clear from the letters that from an early date Thomson's ultimate goal lay elsewhere.

According to Aitken, Thomson was in 1833 introduced by Lady Holland to Lord Melbourne as "the future Professor of Anatomy in the University of Glasgow".[26] Thomson, had, on his mother's side, strong links with Glasgow University, and his father seems also to have encouraged him to look in that direction. In 1839 Sharpey alerted Thomson to the possible vacancy of the Glasgow Anatomy Chair (letter 16). Thomson's designs on this position were until 1848 thwarted by the inconsiderate longevity of the incumbent. It is unclear why Thomson would have preferred the Glasgow Chair to his situation in Edinburgh: Sharpey hinted as much when he advised Thomson in January 1848 to let "Lord John [Russell] know in some way or other that your views are directed towards the Glasgow Chair—He may not know the respective advantages of the position as compared with your present place—which he may very naturally suppose better" (letter 38). The Glasgow school of medicine in the early nineteenth century was in a decrepit condition, and the condition of the anatomy department was among the worst. There is a tantalizing reference in one letter to a list of "reasons" for preferring Glasgow that Thomson had sent Sharpey, but no more (letter 35). It can only be assumed that Thomson saw the Glasgow Chair as having great potential and as a more lucrative prospect than his present position. In 1849 Philip Kelland (1808–79), who held the Chair of

[23] Aitken, op. cit., note 20 above, p. xiii.
[24] Ibid., p. xx.
[25] Ibid., pp. xx–xxi.
[26] Ibid., p. xvi.

Mathematics in Edinburgh, wrote to ask of Thomson's "welfare (I had nearly said prosperity, but that would have come ill from a *poor* Edinburgh Professor to a *rich* Glasgow one)".[27]

Glasgow was to prove Thomson's ultimate professional destination. He held the Chair there for 29 years and (unlike Sharpey) seems not to have been tempted by other openings. During his time in Glasgow, Thomson made a major contribution to the revival of the medical school, greatly increasing the size of the anatomy class.[28] By 1861 the income of the Anatomy Chair was around £750.[29] In addition to his work in his own department, Thomson sought to secure the best appointments to other medical chairs in the University in the face of determined local opposition. These efforts are well-documented in the letters and are discussed more fully below. He was also active in the University's physical renovation: as Chairman of the Buildings Committee, Thomson supervised the transfer of the University to new premises at Gilmorehill between 1863 and 1870.

Upon his retirement in 1877, Thomson came to live with his son in London. There he died seven years later.

There are a number of clear parallels between the careers of Sharpey and Thomson. They were progenitors of a species of medical man that—although well-established in both France and Germany by the third decade of the nineteenth century—was late in developing in Britain. Both were medical academics who for almost their entire careers derived their incomes from teaching and not from practice. In this they differed from previous generations of Scottish medical teachers, and from most of their contemporaries, for whom teaching merely supplemented the practice of medicine. They were not, however, only teachers; they also made original contributions to medical science in the forms of histology and embryology, although this aspect of their activities became less prominent in their later years. In short, their careers reflected the research as well as the teaching role of the medical academic.

A commitment to the development and transmission of medical knowledge went hand-in-hand with a leading role in medical politics. Indeed, the latter activity was seen as vital to the success of the professional and intellectual ideals for which they strived.

Their letters reflect all these preoccupations. They are rich in detail concerning the practicalities of medical teaching, the state of histology, and the politics of professorial appointment in the mid-nineteenth century.

MOUNTING A CAMPAIGN

For almost their entire careers, teaching was Sharpey and Thomson's trade. The correspondence consequently contains much information on the organization and conduct of a course of lectures on anatomy-physiology in nineteenth-century

---

[27] [Philip] Kelland to Allen Thomson, 30 January 1849, Thomson MSS, op. cit., note 1 above, Box 18.
[28] See J. Coutts, *A history of the University of Glasgow: from its foundation in 1451 to 1909*, Glasgow, James Maclehose, 1909, pp. 520–1.
[29] Ibid., p. 576.

Britain. The letters immediately following Sharpey's transfer to London are, in particular, much concerned with business of teaching. They yield numerous insights into the differences, as well as the similarities, the Edinburgh lecturer experienced in his new school.

The first point to stress is that teaching in both Edinburgh and London was, in this period, a business. As an extra-mural teacher Sharpey had been entirely dependent upon student fees for his income, and he and Thomson had had to compete for these fees with the numerous other lecturers in Edinburgh. Sharpey was relatively successful in this competition: in the five years he lectured in Edinburgh the number of students in his class increased four-fold, from 22 to 88. Thomson, in contrast, never secured more than 36 students.[30] In London too, although he was guaranteed a small stipend, his income was mostly derived from fees. Moreover, the competition in London, where numerous schools of medicine existed, was still more fierce. The size of Sharpey's class, and therefore his income, fluctuated violently; in 1838–9 it stood at 374, while in 1863–4 it fell to 91. Student numbers and the rivalry between medical schools is a persistent theme in the correspondence.

The early letters are concerned with the settling of accounts between Sharpey and Thomson following the dissolution of their partnership: indeed, the letter of 2 December 1836 contains an actual account of the expenses and income their school had incurred during the summer of 1835 (letter 7).

Among the assets of the school the chief were the museum and its collection of drawings. Aitken dwelt upon the amount of time Thomson spent, prior to setting up as a teacher, visiting existing museums in Britain and on the Continent to study the preparations they held. From the information derived from these travels he formulated "an extensive list of preparations 'to be made' for teaching purposes".[31] The pains Thomson took to complete this exercise reflect the importance of preparations in the teaching of anatomy and physiology during this period.

In the case of Edinburgh teachers, preparations helped to offset the shortage of cadavers available for dissection. Sharpey and Thomson worked within the framework established by the 1832 Anatomy Act, which had been introduced largely in response to the illegal procurement of bodies that had previously obtained in Edinburgh.[32] Although the Act in theory guaranteed anatomy teachers a regular supply of bodies, in practice difficulties of implementation ensured that supplies remained scarce in Edinburgh. In contrast, Sharpey enjoyed a superabundance of material in London and invited Thomson to use the resources of University College's hospital in preparing materials for his course (letter 12).

Edinburgh teachers had, however, since the eighteenth century made a virtue of the scarcity of dissection material and developed an effective pedagogic strategy based upon the use of drawings and preparations.[33] The Sharpey-Thomson

[30] 'Number of students in Dr Sharpeys Classes in Edinburgh', Thomson MSS, op. cit., note 1 above, Box 16.

[31] Aitken, op. cit., note 20 above, pp. xvi–xviii.

[32] On the background to the passage of the Anatomy Act see Ruth Richardson, *Death, dissection and the destitute*, London, Routledge & Kegan Paul, 1987.

[33] Lawrence, op. cit., note 17 above, p. 195.

correspondence suggests that this tradition remained vital among Edinburgh teachers in the first half of the nineteenth century and that they carried it with them when they moved to other centres.

The museum of Sharpey and Thomson's school remained in Edinburgh when the former moved south. Sharpey made provision for its care until Thomson's return (letter 5). Thomson implicitly acknowledged the sacrifice Sharpey had made when he wrote, on 6 October 1836, "I hope that . . . leaving the preparations will not be inconvenient to you" (letter 6). Upon reviewing the resources available to him in London, Sharpey concluded that the museum at University College, although "showy" (and the work of an Edinburgh man), "is anything but a good working one" (letter 7). A great deal of labour was as a result needed to produce new preparations.

The making of anatomical preparations required great skill; indeed, the finest examples were works of art. Perhaps the most demanding aspect of the mystery was the injection of mercury and other liquids in order to display the distribution of vessels within a specimen. Sharpey delegated this task to a subordinate—possibly one of his demonstrators at the College—and described his efforts to inject the lymphatics at a wide variety of sites. This account suggests that, for its devotees, injection could cease to be merely a means and sometimes became an end in itself. The same passage provides an example of how an exotic specimen (in this case an ostrich) was pressed into service to demonstrate vividly a particular structure less evident in man (letter 7).

Once made, specimens had to be preserved and displayed to optimum effect. A number of letters deal with the liquids best suited to preservation and with the construction of glass cases (letters 18, 19, 21).

Sharpey and Thomson relied heavily upon pictures to illustrate their lectures. Some were reproduced from the standard anatomical works of the time and formed a permanent stock of teaching aids. Sharpey's letter to Thomson of 16 February 1839 concerns the division between them of such pictures (letter 14). Other illustrations were, however, more ephemeral, drawn on a slate or board to accompany a particular lecture: Sharpey employed the services of a "young man" for this purpose (letter 7). A statement of the expenses he incurred under this head has been preserved.[34]

It is clear that Sharpey had relied heavily during his Edinburgh years upon his partner's skills and judgement in the matter of illustrations. Allen Thomson was an exceptionally gifted draughtsman who supplied superb drawings to accompany his own and Sharpey's early publications. After moving to London Sharpey continued to seek Thomson's advice on the materials and methods best suited to producing oil paintings of such anatomical structures as the eye and ear (letters 7 and 8). Sharpey hinted at something of the function that illustrations played in the armamentarium of a teacher: they could serve "to illuminate a dry, at least a tedious part of your course" (letter 7).

[34] See 'Tuson's Account for drawing diagrams and painting', Sharpey-Schäfer Collection, op. cit., note 6 above, ESS/B. 2/19.

Although he deferred to Thomson on questions of illustration, in other aspects of the anatomist's art Sharpey was indubitably the master and Thomson the apprentice. During their partnership Sharpey had taught the anatomical portion of their joint course and Thomson the physiology; Thomson had, however, acted as Sharpey's demonstrator for a time.[35] When Thomson set up in Edinburgh as an independent anatomy lecturer he was obliged turn to Sharpey for guidance.

In May 1837 Sharpey suggested a plan for Thomson's projected course. He also give him the name of a former student who had taken such notes of Sharpey's Edinburgh lectures as might give Thomson further guidance (letter 10). This is an interesting example of the role played by student notes during this period in the perpetuation and dissemination of the unpublished lectures of a teacher. In the following April Sharpey offered Thomson access to the copious supply of human cadavers available to the London teacher to gather teaching materials (letter 12). At a later date the correspondence records Sharpey's dispatch of further specimens required by Thomson (letter 14).

On 24 November 1838, shortly after Thomson began teaching, Sharpey wrote again to congratulate him on attracting more than 20 students. He also took the opportunity to convey some friendly criticism of Thomson's teaching style gleaned from a former student and to give advice on the proper management of dissections. Much of the letter is, however, devoted to one of the perennial technical problems confronting anatomy teachers: the preservation of cadavers. Sharpey had been experimenting with new injections in London and he passed on his results in the form of a recipe (letter 13).

All of the pedagogic techniques so far discussed would have been perfectly familiar to an eighteenth-century anatomy lecturer. In one respect, however, both Sharpey and Thomson were held to be innovators: this was in their use of the microscope as a teaching tool.[36] Both had learned this method during their visits to German medical schools and employed it on their return to Edinburgh; it is likely that the possession of such a novel mode of instruction gave them some advantage in the incessant competition for students between the rival teachers there. Certainly, the Committee which considered Sharpey's application for the London chair held him to have "a great advantage as a teacher in having studied the methods of instruction in the best continental schools".[37]

Thomson claimed that Sharpey's use of a microscope mounted upon a revolving table was "the first attempt made in London to illustrate physiological lectures microscopically".[38] There is some discussion in the letters about the construction of microscopes best adapted to this purpose; Sharpey recommended one design as answering "very well for *exhibition* as intended, for you can by giving the tube a

[35] Thomson, [Life], op. cit., note 3 above, p. 7.

[36] But note Monro's listing of microscopes in his teaching impedimenta: Lawrence, op. cit., note 17 above, p. 199.

[37] 'Report of Committee of the Senate appointed to examine the application and testimonials of candidates for the Professorship of Anatomy and Physiology', University College London [UCL] MSS, AM 1–5 (3), p. 18.

[38] Thomson, op. cit., note 2 above, p. xiv.

screwing motion within the other find the focus with tolerable ease and once found it is not liable to be deranged by the inspector". He went on to say, however, that "It will not be so well adapted to recommend to students as a *working microscope*", which suggests that both he and Thomson were encouraging their students to undertake private microscopic study as early as 1842 (letter 21).

The teaching methods so far described were almost all as applicable to the physiology course Sharpey gave in London as to the anatomy courses he and Thomson had delivered in Edinburgh. The one exception was dissection: at University College descriptive anatomy was taught by another professor and dissections were conducted by demonstrators. This mutuality of techniques is indicative of the considerable overlap between anatomy and physiology that persisted in early nineteenth-century Britain. In France and Germany, physiology had gone far in emancipating itself from its roots in anatomy and acquired an autonomous intellectual and institutional identity, but in Britain this separation was much slower in coming.[39] Physiology's development as a discipline has also been supposed to have been inhibited by the strength of anti-vivisection sentiment in Britain.

The Sharpey-Thomson correspondence offers some support for these claims. Shortly after his arrival in London, Sharpey advised Thomson that his projected course would be "one of physical and physiological Anatomy—to compound small things with great on the plan of the Elementa of Haller" (letter 7). It encompassed general, including microscopical, anatomy, and sought to display the connections between structure and function. His teaching was, in short, in the tradition of *anatomia animata*.

So ambiguous was the identity of Sharpey's subject that some of his students when entering on further study at other universities claimed attendance on his course as a qualification in physiology, and others in anatomy. The Dean of Medicine in Edinburgh was obliged in 1838 to write to the authorities at University College seeking clarification.[40]

Taylor has, moreover, found evidence that Sharpey shared the revulsion of many of his countrymen to the experiments on animals conducted by François Magendie—the archetypal vivisectionist—some of which he had witnessed during his visit to Paris.[41] However, he has also pointed out that this does not indicate a total rejection of vivisection: Sharpey merely objected to the poor design of Magendie's experiments and the *unnecessary* suffering they caused. Moreover, Sharpey conducted experiments on animals both for his own benefit, and for his class.[42]

Sharpey's first letter (2) in the collection contains an account of a vivisection he performed in Edinburgh in conjunction with a number of colleagues, including

[39] See G. L. Geison, 'Social and institutional factors in the stagnancy of English physiology, 1840–1870', *Bull. Hist. Med.*, 1972, **46**: 30–58.

[40] Thomas Stewart Traill to the Secretary of University College London, 11 November 1838, UCL MSS, College Correspondence, 1838 Oct./Nov., 4376–4399 (4394).

[41] Taylor, op. cit., note 2 above, pp. 151–2, 255.

[42] Ibid., pp. 140–4.

Robert Christison. Other letters show Sharpey practising electro-physiology and experimental toxicology (letter 9, 33, 59). The correspondence also confirms Sharpey's use of experiment as a teaching technique: in December 1837 he informed Thomson, "I have continued to show a few experiments (more on *dead* than on *living* animals however)" (letter 11). Such use of experimental demonstrations in lectures represented, moreover, no departure from the pedagogic tradition in which Sharpey had been trained: in the eighteenth century the Monros had made regular use of experiments in their teaching.[43]

If there were no intellectual or ethical obstacles to Sharpey's use of live animals in his teaching, he did face certain practical hindrances. In particular, the *supply* of experimental subjects proved a problem in London. In December 1836 he lamented, "I can get nothing on a few hours notice, and the distance to the rabbit market, the slaughter house &c. are very distressing—I think I must send to Edinburgh for frogs" (letter 7). Although his new base was better provided with materials for human anatomy, Edinburgh was evidently a more convenient place for the experimental physiologist. Twenty-one years later these problems had still not been solved; and Sharpey proposed to overcome them by breeding his own rabbits and frogs (letter 60).

It would therefore be a mistake to view Sharpey's teaching merely as a species of animated anatomy. Although the course undoubtedly leaned towards a morphological approach to questions of function, this was partially offset by his awareness of the importance of experimental work. There is no question that Sharpey was fully conversant with both the results and with the methods of contemporary Continental physiologists and sought to acquaint his students with these researches.

In one department, however, he freely admitted his own shortcomings. In 1851 he concluded a discussion of recent work on the physiology of digestion by declaring: "I wish I knew more of chemistry!" (letter 47). In the following year he attempted to remedy this lack by enrolling as a student of practical chemistry at the Birkbeck Laboratory (letter 49). It should be noted that his professed lack of chemical understanding did not prevent Sharpey from expounding and demonstrating to his students Claude Bernard's recently-published researches on the glycogenic function of the liver (letter 47).

Despite his interest in the latest trends in experimental physiology, Sharpey had no claims to be an original worker in this field. Similarly Thomson was always more a morphologist than a student of function. One subject area in which they could, however, claim the status of original researchers was in the field of microscopic anatomy.

HISTOLOGY

Sharpey and Thomson's precocious use of the new achromatic microscope has already been noted. They formed part of a group of Edinburgh workers, which also included John Goodsir, Martin Barry, and John Hughes Bennett, who were active

---

[43] Lawrence, op. cit., note 17 above, especially pp. 206–8.

microscopic researchers in the years immediately preceding and following the promulgation of the cell theory in 1838. Sharpey himself disliked the term "histology", which came into general usage in Britain in the mid-nineteenth century, preferring to describe these endeavours as "microscopical anatomy" (letter 75).

The correspondence provides a number of insights into how Sharpey and Thomson viewed other members of this small group of histological pioneers. In May 1845 Sharpey discussed a collection of Goodsir's papers he had received. He doubted the wisdom of reprinting some of these researches, maintaining that "the paper on centres of nutrition was well as a passing contribution, but scarcely deserves its present place." This comment shows how widely contemporary and retrospective judgements can diverge: Goodsir's paper on 'Centres of nutrition' is now considered to be his most significant contribution, foreshadowing some aspects of Virchow's later work. Sharpey's remarks also illuminate his own attitude to science: he chided Goodsir for being "too anxious to gain a reputation as a generalizer in science . . . the great aim no doubt of all science but not to be done rashly" (letter 22). Contemporaries sometimes faulted both Sharpey and Thomson for their reluctance to attempt syntheses of the particular observations they had made. However, Sharpey's aversion to premature generalization had its roots less in naïve empiricism than in a venerable Scottish methodological tradition, which he would have imbibed in his early university days.[44]

His strictures about particular papers notwithstanding, Sharpey was in no doubt that Goodsir was "an excellent observer and sound headed man". His opinion of the work of Martin Barry, another Edinburgh man who removed to London, was altogether more harsh. In a long letter written in October 1845 Sharpey launched a devastating attack upon the validity of Barry's observations and upon his scientific integrity. These remarks were occasioned when Thomson asked for Sharpey's opinion about Barry's claim that muscle fibres were composed of a "double spiral" of filaments. Although controversial, this theory had received the support of others, including the influential Richard Owen.[45] Barry had even persuaded Sharpey himself of the truth of his doctrine. However, upon repeating these observations at his leisure, Sharpey became convinced that Barry's doctrine was the result merely of an optical illusion (letter 24).

Sharpey went on to cast doubts upon other aspects of Barry's work, and, in particular, his claim that the blood corpuscles were capable of spontaneous movement. Sharpey proved to his own satisfaction that the motion Barry had observed was the result of the action of cilia. Although he couched it purely in terms of a contrast between accurate and erroneous observations, important theoretical considerations were implicated in this issue. The spontaneous motion of blood corpuscles could be adduced as evidence for the existence of a special "vital force": indeed, W. B. Carpenter drew just this inference. Sharpey, on the other hand, was

---

[44] See G. N. Cantor, 'Henry Brougham and the Scottish methodological tradition', *Stud. Hist. Philos. Sci.*, 1971, 69–89.

[45] See R. Owen, *The life of Richard Owen: by his grandson the Rev. Richard Owen, M.A.*, 2 vols., London, John Murray, 1894, vol. 1, p. 200.

opposed to any such doctrine: as early as 1831, he invoked ciliary motion to disprove similar claims made by another researcher.[46]

More generally, Sharpey's remarks supply insights into the nature of contemporary conflicts over the interpretation of microscopic observations. These early histologists were confronted with the formidable task of *conceptualizing* what was, in effect, a new world. Precisely because they were pioneers, they could rely upon no established interpretative framework.[47] Under these circumstances disagreement was inevitable. Each researcher brought particular preconceptions and sensitivities to his investigations. Sharpey was "startled" by Barry's evident failure to see the phenomena of ciliary motion which was so obvious to him; but *his* eye had been sensitized by long years of searching for such cilia, whereas Barry was looking for something quite different.

Such conceptual conflicts could only be overcome when a *social* consensus was established within the community of microscopic observers. Sharpey's letter exemplifies attempts to exert social control of this kind upon Barry. In both instances of disagreement Sharpey repeated Barry's observations in his presence and in the company of other microscopists. In the latter case Barry appeared to succumb to this pressure: "in the end he turns to Marshall and says, that if he was to be corrected he was very happy to be put right by Dr Sharpey." What provoked Sharpey's particular wrath and contempt was Barry's subsequent profession of the same opinions despite his ostensible submission to the judgement of his peers. His case provided "a fearful spectacle of morbid craving for scientific distinction" (letter 24). Barry's offence was, therefore, as much against the *morality* as the technical standards of science.

Some of the disagreements between histologists of the 1840s may be ascribed to the fact that they for the most part employed fresh, untreated tissue without the use of the staining agents and fixative techniques that later became available. A letter (77) of 1865 reveals, however, that the interpretation of treated specimens could be equally problematic. It also suggests that as a Secretary of the Royal Society with responsibility for the refereeing of histological papers, Sharpey played a major role in adjudicating such questions.

MEDICAL CHAIRS

The final notable theme of these letters is the politics of professorial appointment in nineteenth-century Britain. During the course of the correspondence Sharpey and Thomson were involved in several such contests, sometimes as mere spectators, but more often as active partisans.

The first of these episodes was the prelude to Sharpey's own appointment in 1836 to the Chair of Anatomy and Physiology at the University of London. This appointment has been considered by Mazumdar in the context of the wider goals of the University's founders. She has seen Sharpey's success as reflecting the bias of the

---

[46] See Taylor, op. cit., note 2 above, p. 146.

[47] I have discussed this point more fully in 'John Goodsir and the making of cellular reality', *J. Hist. Biol.*, 1983, **16**: 75–99.

dominant group on the College Council towards Edinburgh-trained men and against the products of the London anatomy schools.[48]

The correspondence adds a few additional details to existing knowledge about Sharpey's appointment. It shows that Robert Carswell, a long-time acquaintance of the Thomson family, furthered Sharpey's cause (letter 1). Carswell is the probable addressee of the testimonial for Sharpey that Thomson composed; parts of this document were embodied in the report that the appointments committee eventually produced (letter 3). Richard Quain's role in going to Edinburgh to hear Sharpey lecture is recorded in other documents;[49] the letter of 18 July 1836 makes it possible, however, to place an approximate date on this visit. It reveals too, that Quain heard a lecture by another contender, Alexander Lizars, which suggests that the selection committee had not yet made up its mind between these two candidates.

The same letter shows Sharpey, in conjunction with the Thomsons, utilizing the strong links between Edinburgh and the London University to advance his claims. But there is also evidence of hostility among some members of the College Council to his claims: Henry Warburton, in particular, appears to have opposed Sharpey's appointment (letter 2). Warburton and at least one other of the Council were for Richard Grainger, a local candidate.[50] Mazumdar's assumption that there was a bias towards Edinburgh men among the Council members must, therefore, be qualified.

Thomson's long-standing interest in the Glasgow Chair of Anatomy has already been mentioned. When in March 1839 Sharpey erroneously alerted him to the possibility of a vacancy there, he also broached the subject of how best to secure Thomson's appointment. In this letter, two themes that attain great prominence later in the correspondence emerge. The first is the connection between academic and national politics. Because the Glasgow post was a Regius Chair it lay in the gift of the Crown. Sharpey therefore advised Thomson to use James Clark, Physician-in-Ordinary to the Queen, as a referee if required to do so by Lord John Russell, the Home Secretary (letter 16). His family's close links since the late eighteenth century with Whig politicians, and the Russell family in particular, were to stand Thomson in good stead in such attempts to secure patronage.

But long before James Jeffray made it possible for these manoeuvres to come to fruition, another opportunity opened up for Thomson. In June 1842 Sharpey advised him that W. P. Alison, Professor of the Institutes of Medicine, was (in the time-honoured manner) likely to transfer to the Edinburgh Chair of the Practice of Medicine, thus creating a vacancy in the medical faculty. In this case, patronage was in local hands, for appointments to most Edinburgh University chairs were still vested in the Town Council. The fierce contest that preceded J. Y. Simpson's appointment to the Midwifery Chair in 1839 reveals the scope for factional conflict

---

[48] P. M. H. Mazumdar, 'Anatomical physiology and the reform of medical education, 1825–35', *Bull. Hist. Med.*, 1983, **57**: 230–46, especially pp. 242–4.

[49] See Taylor, op. cit., note 2 above, p. 137.

[50] When the Council came to make its appointment, Warburton read out two letters in support of Grainger's candidacy. See University College Records, Session of Council, Thursday 11 August 1836.

that professorial contests in Edinburgh could provoke.[51] Hence Sharpey's warning to Thomson against being "taken up . . . by a particular party". In the event, Thomson's appointment to the Institutes Chair was apparently uncontentious; all that was required was the enlistment of an impressive group of referees (letter 20).

Once installed in the Edinburgh medical faculty, Thomson was ideally placed to facilitate Sharpey's return to Edinburgh in 1846 as Professor of Anatomy. He alerted him to the imminent vacancy in December 1845. In his reply Sharpey, after reviewing the pros and cons, indicated his willingness to return to Edinburgh if Thomson could guarantee an uncontested election; once again he revealed his dislike for and fear of the potential acrimony that the contest for an Edinburgh chair could generate. With the aid of James Syme and, perhaps, J. Y. Simpson, Thomson seems to have had no trouble in meeting this requirement; and only the blandishments of Sharpey's London colleagues ultimately thwarted the appointment.

In Glasgow a more diverse range of interests bore upon professorial appointments. This was the result both of the greater number of Regius Chairs and the different relationship between the University and the local and national polity. When in 1847 the Glasgow Anatomy Chair finally became vacant, Sharpey briefed Thomson on his likely rivals and the patronage they might enjoy. The Prime Minister, the Home Secretary, and the Lord Advocate emerge as the most important players; but the possibility of the intervention of the Glasgow MPs is also mentioned. The same dramatis personae were to figure in subsequent Glasgow contests. The Lord Advocate's role deserves special notice. After 1765 this officer had developed into a veritable minister for Scotland with formidable executive powers. The Lord Advocate also acted as a patronage broker for the government north of the border; appointments to Regius Chairs formed part of this political function.[52] Sharpey had no doubts that, with his political connections, Thomson was assured of success; it was nonetheless essential to avoid complacency and to ensure that "there is no failure through mismanagement" (letter 35).

In order to realize Thomson's potential support, it was necessary to communicate his claims to those in power. This appears to have been no straightforward process. Even for a member of the Thomson family, direct access to someone like Lord John Russell appears to have been difficult: it was necessary to employ some intercessor. In January 1848 *Lady* John was mentioned as someone who might perform this role (letter 38). She appears to have been on familiar terms with Thomson, consulting him in 1849 over the choice of a tutor for her son (letter 44). A more important intermediary, however, was a politician's "medical confessor": that is, the practitioner who attended on him, and who might have some call upon his attention. In particular, royal physicians and surgeons, such as James Clark and Benjamin Brodie, seem to have enjoyed privileged access to the mighty (letter 39).

Such metropolitan influence needed to be complemented with political support within Scotland; and here the role of the Lord Advocate was crucial. When his

---

[51] See J. Duns, *Memoir of Sir James Y. Simpson, Bart.*, Edinburgh, Edmondston and Douglas, 1873, pp. 98–103.
[52] See A. Murdoch, *The people above: politics and administration in mid-eighteenth century Scotland*, Edinburgh, John Donald, 1980, especially pp. 13–14.

appointment to the Glasgow chair was assured, Thomson wrote a fulsome letter of thanks to Andrew Rutherfurd, the Lord Advocate. He took care to invoke the memory of John Thomson, who had moved in the same Edinburgh Whig circles as Rutherfurd, maintaining that "nothing concerning me could have given greater satisfaction to my father than my receiving this appointment, to which he from an early period encouraged me to look forward."[53] William Thomson gave further proof of his family's gratitude by supporting Rutherfurd's candidacy as Rector of Glasgow University later in 1848.[54] Patronage had its price.

The professoriate of Glasgow University formed a notoriously conservative, self-perpetuating clique. Despite his Millar blood, Thomson was to some degree viewed as an intruder: he wrote, perhaps facetiously, in 1852 that some of his medical colleagues had threatened to resign upon his appointment (letter 48). Thomson saw his role in the University as that of a reformer implacably opposed to the—in his eyes—corrupt mode of filling chairs that had hitherto prevailed.

His first challenge came in 1852 when the question of a successor to Thomas Thomson (no relation) in the Chair of Chemistry arose. The latter had attempted to ensure that his chair remained in the family by employing his nephew as an assistant lecturer for some years prior to his death; such nepotism was not uncommon in Scottish universities. Upon the old man's death, the medical faculty drew up a testimonial in favour of the nephew in order to forestall the appointment of an outsider. There were precedents for this stratagem in earlier efforts by the medical faculty to secure a chair for an inside candidate by presenting a common front to the Crown.[55] Thomson, however, refused to participate in this exercise, seeing it as a manifestation of the wretched parochialism and "Edinophobia" of the Glasgow professoriate. Wider political interests were, however, also involved. Thomson noted that the emergence of a Conservative government had brought with it an attempt to revert to "the old tory way of making appointments through the Duke of Montrose" in the University.[56]

Thomson's response was to turn to Sharpey to find some outside contender to resist the claims of the internal candidate. In particular, he hoped that Thomas Graham, the Professor of Chemistry at University College, might be induced to offer himself for the Glasgow Chair (letter 48). This was a shrewd choice since Graham was not only a respected chemist, but also originally a Glasgow man. Sharpey's reply was, however, discouraging, pointing out that Graham's situation in London was too comfortable for him to contemplate a move. He did mention a number of other candidates who might be suitable, including the Edinburgh lecturer Thomas Anderson who finally succeeded to the Chair. In this letter Sharpey revealed the

---

[53] Allen Thomson to Andrew, Lord Rutherfurd, 9 February 1848, National Library of Scotland [NLS] MS 9714 ff. 178–9. On Rutherfurd see G. W. T. Omond, *The Lord Advocates of Scotland: second series 1834–1880*, London, Andrew Melrose, 1914, pp. 47–125.

[54] William Thomson to Andrew, Lord Rutherfurd, 14 November 1848, NLS MS 9714 ff. 300–301.

[55] Coutts, op. cit., note 28, above, pp. 499–500.

[56] On the role of the Duke of Montrose in Glasgow University patronage see A. Duncan, *Memorials of the Faculty of Physicians and Surgeons of Glasgow 1599–1850: with a sketch of the rise of the Glasgow Medical School and of the medical profession in the west of Scotland*, Glasgow, James Maclehose, 1896, p. 174.

importance of "friendship" in influencing the support a candidate might secure (letter 49). "Friendship" in this sense referred to the complex web of personal relations and obligations in which potential patrons were enmeshed. An individual's attitudes and loyalties might be as much influenced by such commitments and loyalties as by party or institutional ties.

This episode adumbrated many of the features of the far fiercer contest in 1859–60 for the Glasgow Chair of Surgery. There was the same interaction between metropolitan and local influences, and between the micro- and macro-political. In this case, however, the Glasgow Members of Parliament, who figured only slightly in 1852, loomed large. The role played by MPs as a channel of communication between important interests in their constituencies and the London government has been described by Bourne in his recent study of nineteenth-century patronage.[57] Yet the Members were subject to a variety of sometimes contradictory demands; loyalty to party often mattered less than the need to appease local factions and the demands of "friendship".[58]

The correspondence makes it clear that appointments to Regius Chairs were subject to similar pressures. Thomson in November 1859 declared it to be "a great grievance that the appointments to scientific chairs in the Universities should be influenced in the manner which it appears is being done by the Glasgow Members in the case of the Surgery Chair." The MPs in question, Walter Buchanan and Robert Dalglish, prided themselves on being "sapient radicals", but they had "taken up the cause of the Candidate of the greatest & most unscrupulous Tory connection in Glasgow on the grounds of private friendship and the flimsy & absurd view that Surgeons of Glasgow growth should alone obtain places in its University 'Our own fish guts &c.'." The candidate in question, G. H. B. Macleod, had been sponsored by the incumbent of the Chair in much the same way as Thomas Thomson had taken up his nephew. Lawrie had persuaded Andrew Buchanan, a traditionalist member of the medical faculty, also to support Macleod; and Buchanan had, in turn, prevailed upon his brother Walter to use his influence in the cause. Dalglish had been recruited through another friendship network: he was a neighbour of Macleod's father, a prominent Glasgow clergyman (letter 63).

Macleod therefore had formidable backing among the tight-knit oligarchy that dominated Glasgow. These saw the appointment merely as a means of gratifying local interests and of reinforcing civic chauvinism. Against this, Thomson asserted that the appointment should be determined on the basis of merit, and, in particular, the scientific credentials of candidates. There are clear overtones here of the radical critique of the "Old Corruption" which had been current in British political discourse since the eighteenth century; but Thomson's ethos also mirrors that of nineteenth-century campaigns to fill posts in the public service on the grounds of candidates' educational and professional qualifications, rather than on the basis of

[57] J. M. Bourne, *Patronage and society in nineteenth-century England*, London, Edward Arnold, 1986, pp. 153–4.
[58] Ibid., pp. 137–8.

the "interest" they could command.[59] By this standard there was, he maintained, no question but that the post should go to Joseph Lister.

The irony was that Thomson could only hope to attain this goal by means of his own patronage network; it was to this end that he wrote to Sharpey on 27 November 1859 (letter 64). As in the case of his own appointment to Glasgow, Thomson could count upon the support of both the Home Secretary and the Lord Advocate. Such was the pressure from the Glasgow Members, however, that the former hesitated to appoint Lister (letter 65). Thomson sought to counteract such political pressure by appealing to the authority of senior metropolitan medical practitioners like Brodie and Clark. Implicit in his argument is the principle that only professional men were capable of making a judgement in these matters: "A Member of Parliament is not a better judge of the fitness of professors of surgery than other men." (letter 66).

Thomson pursued this strategy in the early weeks of 1860. He wrote on 16 January to Brodie who, as someone interested "in the advancement of scientific instruction in medicine", might intercede with the Home Secretary on Lister's behalf (letter 68).[60] Sharpey reported that Brodie was, however, reluctant to intervene and that the Home Secretary seemed about to succumb to local pressure (letter 69). Two days later Sharpey sent a still more pessimistic note complaining that the Home Secretary was about to make "the interests of the University and of Medical Education in Scotland subordinate to the gratification of a political supporter" (letter 70).

Sharpey had, however, misjudged the position. The very intensity of the politicking for Macleod appears to have acted against him. In particular, the Senate of Glasgow University felt its own authority threatened by the interest outside factions had taken in the Surgery Chair and protested to the Crown against this interference.[61] On 28 January 1860 Lister was appointed Professor of Surgery in Glasgow University.[62] There is a gap in the correspondence at this point so Sharpey and Thomson's reaction to this success is not known. The next letter in the sequence was written after Lister had begun to teach in Glasgow; it describes his success and the likely prospect that he would soon obtain a surgeoncy at the Royal Infirmary (letter 71). It was during his Glasgow period that Lister began his experiments in antiseptic surgery.

Although no subsequent event matched the intensity of the contest for the Glasgow Chair of Surgery, two sequels do figure in the correspondence. The first of these occurred in 1875 when the Physiology Chair in Glasgow became vacant; the second two years later when the question of Thomson's own successor arose (letters 93–96, 98–99). A number of familiar themes are reprised in these letters: in both cases the "best" (in Sharpey and Thomson's estimation) candidate had to contend against a rival with local influence. In the case of the Physiology Chair, influential

[59] Ibid., pp. 166–76.
[60] Lister had already applied directly to Brodie for support, at Sharpey's suggestion. See Joseph Lister to William Sharpey, 8 October [1859], Edinburgh University Library MSS, AAF; Lister to Sharpey, 13 October 1859, NLS MS 9814 f. 141.
[61] See Coutts, op. cit., note 28 above, p. 582.
[62] See R. B. Fisher, *Joseph Lister 1827–1912*, London, Macdonald and Jane's, 1977, pp. 96–7.

professional opinion was again mobilized to assure the success of the favoured candidate. The case of the Anatomy Chair was less easy to manage; it transpired that even the Queen's Physician could express an opinion only when asked to do so by the Home Secretary (letter 98). Nonetheless, the eventual outcome of this contest too was satisfactory to Sharpey and Thomson.

Sharpey and Thomson clearly saw themselves as in the intellectual vanguard of Victorian medicine. Their self-appointed mission was to ensure that merit, and specifically scientific merit, triumphed above all other considerations in competition for academic posts. They set themselves against all forms of nepotism and jobbery. Nor was this campaign confined exclusively to the Universities: in 1854 they intervened to ensure that the "best" man obtained an Assistant Physicianship at St Bartholomew's Hospital against "the son of one of the Old Physicians of the Establishment . . . [who] carries with him all the influence among the Governors which nepotism . . . can command" (letter 50). In this as in the other cases discussed, they insisted on the right of professional control over appointments against the claims of all lay interests.

Editorial Note

The letters are arranged in chronological sequence. The layout follows that of the original letters as closely as possible. The editor has preserved the spelling and punctuation of the original letters.

Editorial interpolations are inserted between square brackets.

[...] indicates a word that is illegible. Sharpey sometimes used dashes to end sentences; and these have been rendered by the longer dashes in the text.

1

15th July 1836

My dear Sharpey,

The day before yesterday I got a summons from Carswell[1] to see him in some way or other as soon as possible and when I called on him I found you were the subject of conversation. From what he said I think there is every likelihood of your being chosen by the Council of the London University, but of course we can have but a very imperfect notion at present. The Faculty of Medical professors were to meet today to discuss the matter & Carswell says that their suggestion is very generally attended to by the Council. Mayo[2] seemed to have an inclination to offer himself and Grainger[3] is I suppose a candidate but Carswell seems to think that others as well as he is himself will be for having you.

I need not say what tumult of feelings all this raised in my mind.

The emoluments of the situation appear to be nearly £800 a year and you will deliver your lectures in much more advantageous circumstances there than you can do in Edinburgh. [I]t increases your chances of the Edinburgh chair[4] and of preferment in every way and as far as I can see is most advantageous for you.

I wrote you thus promptly in case it were possible for me to be of any use to you in this matter. Write me at all events and let me know what you are thinking of the matter.

I weary to hear something of Surgeons Square.[5] Indeed I feel melancholy whenever I think of it. & what will it be when you are away also?

Yours ever,
Allen Thomson

Address to me care of John Murray Esq., 50 Albermarle Street if you don't put your letter under cover to the Duke [of Bedford].[6]

If I get any information I shall write you again soon.

My papers will be down next week[.] You may have as many as you like to give away.

---

[1] Robert Carswell (1793–1857), in 1836 Professor of Pathological Anatomy at University College London. An intimate of the Thomson family, he had been employed by John Thomson in the 1820s to make a collection of illustrations of morbid anatomy. For this purpose he visited the Continent with William Thomson, and was in Paris at the same time as Allen Thomson (1828–9).

[2] Herbert Mayo (1796–1852), Surgeon at the Middlesex Hospital and Professor of Anatomy at King's College, London since 1830. He became Professor of Physiology and Pathology there in 1836.

[3] Richard Dugard Grainger (1801–65), Lecturer in Anatomy at the Webb Street School which he had inherited from his brother Edward.

[4] I.e., the Chair of Anatomy at Edinburgh University.

[5] Surgeon's Square, Edinburgh: the site of Sharpey and Thomson's and other extramural teaching establishments.

[6] See note 5.5, below.

2

Edinburgh 18th July, 1836

My Dear Thomson,

Many thanks for your kind letter. I am quite alive to the nature of its contents, but I assure you that when I received it you were better aware of what was going on than I was. Mr. R. Quain[1] heard a lecture from each of the Teachers[2] here without their knowing of it, he called on me after and spoke of the change about to take place asking me at the same time if I would be likely to come forward as a candidate. I informed him that I would relish teaching the branches he mentioned in a chair of the Lond. University very greatly indeed. I met him afterwards at dinner but no more passed between us on the subject, and your letter which I got yesterday was the next intelligence I had. This afternoon I had one from Mr. Quain advising me to come up to London as early as convenient. I called on your brother,[3] and the result is that I start for London on Saturday. I can easily do this as the Session ends next week.

Dr. W[illiam Thomson] bids me request you to go over the list of Council with Mr. James Mylne[4] – lest he know any one he could come at, he also spoke of Mr. J. Murray[5] knowing Mr. Greenough[6] who is one of the council. I fear many of them may be inclined to listen to Somerville[7] and there is one whom I certainly could not expect to look favourably on me. I need not say I mean W—n [Warburton].[8]

I would say a great deal more to you my Dear T. on this matter, and what plans in the event of success we might pursue, but I feel that the prospect is still too distant to permit me (naturally the reverse of sanguine in my disposition) to indulge in what might turn out to be day dreams.

To speak of things as they are, and return to Surgeon's Square. We have got a splendid skeleton of one porpoise and most successful injection of the *wonderful net* in another. I have also made a good injected preparation of a turtle heart. — Mr Seaton[9] and I tried Poiseuille's Exp'[10] on a dog much exhausted by having had his gullet tied a couple of days before the pipes were put into the carotid & crural, and the thing did well, the perfect equality of pressure is no exaggeration, it was quite conspicuous except when once or twice notwithstanding our constant [. . .] the tubes got slightly obstructed. The pressure was not great owing to the weakness of the dog. I have ordered the stopcocks to be taken off and made moveable, so that they may be fixed into the tubes, for we required corks in our experiments to plug the [. . .] in one vessel till we had got one fixed into another. I suggested to Christison[11] that we should make trial with it of the force of the heart under the influence of different poisons &c. and we mean to do it – Seaton will give you the results he is the note taker.

Sandy Monro[12] was married the other day, & Douglas Maclagan[13] becomes folded up tomorrow. There has been a regatta where 99 of the 100 spectators knew nothing of what was passing. Your paper is highly approved of. It is really a good thing after all to be forced to write in the Cyclopaedia.[14] By the way I praise God that Echinodermata is at last packed up with Messrs Sherwoods address on the back of it for the mail tomorrow. I was a fool not to have availed myself of your [. . .] to help me with the drawings. I have been sadly at a loss about them. Some are from Tiedemann[15] copied by Mr. [. . .] Scott. – Some from [. . .] by Mr Boyd and by myself. The engravers will have the pleasure and privilege of reducing them.

I would say more but till the London affair is over, one way or other, I shall feel in an uncomfortable state.

I am my dear Thomson

Yours most sincerely

W. Sharpey

[1] Richard Quain (1800–87), Professor of Descriptive Anatomy at University College London since 1832.

[2] Alexander Jardine Lizars, another private anatomy teacher in Edinburgh, was also a candidate for the University College Chair ('Report of Committee of the Senate appointed to examine the Applications and Testimonials of Candidates for the Professorship of Anatomy and Physiology', University College London MSS, AM 1–5 (3), p. 6.) It was probably his lectures that Quain also attended during his visit to Edinburgh.

[3] I.e., William Thomson (1802–52), Allen's half-brother, in 1836 a private lecturer in the Practice of Physics in Edinburgh.

[4] James Mylne was a member of the Council of University College London from 1830 to 1840.

[5] Probably John Murray (1808–92), the son of the publisher whose home Thomson was then using as a mailing address. Both the elder and younger Murray had close ties with Edinburgh; the latter had been Allen Thomson's schoolmate.

[6] George Bellas Greenough (1776–1855), geographer and geologist, was a member of the original Council of the London University.

[7] Presumably William Somerville (1771–1860), Physician to the Chelsea Hospital.

[8] Henry Warburton (1784?–1858), MP and founder member of the Council of the University of London.

[9] Possibly, Edward Cator Seaton (1815–80), an Edinburgh medical student who graduated in 1837.

[10] Jean Léonard Marie Poiseuille (1797–1869) a French physiologist who in 1828 published reports of experiments designed to measure the arterial blood pressure: 'Recherches sur la force du coeur aortique', *Arch. gén. Méd.*, 1828, **8**: 550–4.

[11] Robert Christison (1797–1882), Professor of Materia Medica and Therapeutics in Edinburgh.

[12] Alexander Monro III (1773–1859), Professor of Anatomy in Edinburgh. In 1836 he married for the second time.

[13] Presumably Andrew Douglas Maclagan (1812–1900), who graduated MD at Edinburgh in 1833 and became a Surgeon at the Royal Infirmary prior to commencing lectures on materia medica in 1839.

[14] I.e., the *Cyclopaedia of anatomy and physiology*, 5 vols., London, Longman, 1835–59, edited by Robert Bentley Todd; see note 6.1 below.

[15] Friedrich Tiedemann (1781–1861), a German physiologist who was the author of several anatomical atlases.

3

Campden Hill

[July 1836]

My dear Sir,

In answer to your inquiries concerning my friend Dr. Sharpey, I can have no hesitation in expressing the very high opinion I entertain of his merits.

I have now been on a footing of great intimacy with Dr Sharpey for some years. During that time his amiable moral qualities have made me sincerely attached to him and I have had the happiness of seeing him respected and lauded by all who have the pleasure of his acquaintance and friendship. I have had the best opportunity of knowing that Dr Sharpey in addition to excellent abilities is possessed of a remarkably sound judgement and very extensive information of a professional and general kind.

3

Dr. S. turned his attention in an especial manner to Anatomy at an early period of his professional career. He is deeply versed in the literature of this department of medicine. He is universally admitted to be a clear and accurate demonstrator of the parts of the animal body. He has a competent knowledge of comparative anatomy and is very intimately acquainted with minute or general anatomy — Indeed I believe there are few who surpass him for the knowledge of this branch of the science. His knowledge of physiology is also extensive altho. he has not delivered lectures on that subject.[1]

He takes a great interest in the study & investigation of everything relating to Anatomy is a careful and accurate observer of renown to medical men of every thing relating to Anatomy and may in every point of view be regarded as a truly scientific Anatomist.

Dr. Sharpey has had the advantages of an excellent education. He had seen so much of the world as to have acquired that freedom from prejudice which travelling generally gives. In his visits to the most celebrated Continental Medical Schools he has made himself well acquainted with the mode of teaching Anatomy and other branches of medicine pursued in them.

Dr. S. has with hard labour formed a considerable Anatomical class and were he to remain there would soon I doubt not take the first place as an anatomical teacher. He delivers his Lectures in a clear and attractive style, and is in the habit of making use of drawings and models for the purpose of assisting the illustration. He is very punctual,[2] very attentive to the students, mild and amiable in his demeanour, straightforward gentlemanly and honourable in his conduct so that were he to become professor in your University he could not fail there to add to the reputation of the school to insure the attachment of the students & the respect and esteem of his colleagues as he has already done in Edinburgh.

So high an opinion indeed have I of Dr Sharpey that I may say you will do well to caution me against allowing the expression of it to be moderated by the desire I naturally feel to retain him in Edin. For there is no person companion [*sic*] whose loss I shall more feel as a friend and coassistor in labours. There is no one in my humble opinion who ⟨My opinion then is disinterested when I say that should you obtain Dr Sharpey's services at the London University there is no one who⟩[3] as a sound headed and scientific man and as a teacher in his particular department should you obtain his services will be a greater blank in the Edinburgh school and a greater acquisition to the London University.

You have asked me for my opinion and I have given it thus freely and perhaps presumptuously but I believe I may say that there are many others who entertain an equally favorable opinion of Dr. Sharpey that there are few who have had an equally good opportunity of judging him.

---

[1] This account of Sharpey's virtues is closely paraphrased in the Committee's 'Report', op. cit., note 2.2 above, p. 17.

[2] This statement is quoted ibid., p. 19.

[3] Thomson intended the words in the angle brackets, written after this paragraph, to be inserted here.

4

(London 1836 [The date is in Thomson's hand])

My dear Thomson,

I was not in time for the post.

I was elected today Professor of Anatomy, subject to such regulations as the Council may deem expedient.

The condition attached is merely this. I have been elected to fill the vacancy occasioned by Dr. Q's[1] resignation, in short into his place, but the contemplated and wished for alterations of duties &c. have not yet been adopted, it was felt this would delay the matter and the present proceedings makes all right. It was fortunate that the discussion respecting the allotment of the duties was not entered on, as time would have been lost and perhaps another adjournment been the consequence.

There was a division in the council but this is private. I had an immense majority. 9 to 2. Don't mention this.

My great desire now is to do justice to the appointment.

I will leave this Saturday. I think by the Dundee Steamer. I will be a couple of days in Arbroath and see you in Edin. on Thursday evening.

This is my present intention but I may change it.

Yours most sincerely,
W. Sharpey

[1]I.e., Jones Quain (1796–1865), Professor of General Anatomy at London University from 1831 to 1835.

5

Edinburgh 3rd September 1836

My dear Thomson,

I write in a desperate hurry, and indeed it matters little for I have little to say.

Andrew[1] is to inspect the museum and put on a fire occasionally, during the winter, under Mr. Ramsay's[2] superintendence, whom I have seen and conversed with on the subject.

Macdonald[3] is in Edinburgh again, he will not bother you or me about the museum, I have settled that. I think he is not positively certain of commencing but the probability seems to me in favour of his doing so.

I have carefully avoided directing the students as to their future teacher, but I can see from conversing with several that Handyside[4] will be their preference, he (Handyside) is not in Edinb.

Your friends here think it advisable for you to announce your intention of teaching anatomy next year, and your brother and I have concocted an advertisement subject to your approval to be put in the papers when the proper time arrives. Of course you will let him know your opinion on it.

I start in a few hours by the steam boat for the new scene of my labours. I hope it may be as happy and comfortable a one as that which I leave.

I wrote about the plants & presume they will be sent for the Duke's[7] inspection.

With sincerest wishes for your welfare I ever am

My dear Thomson

affectionately yours

W Sharpey

Dr Allen Thomson

(The Doune of Rothiemunhar [in Thomson's hand])

[On the reverse of this letter, in Thomson's hand:]

Sharpey.

His new situation. his modesty at first. his confidence in himself. Desertion of Surgeons Square.

My plans to lecture on Anatomy. Preparation for this. His note of Lectures & plan of course. his advice.

My hopes of being with him in spring.

Note to Imlach[8]

Separate testimon.

note about Classroom & answer to his letter.

William

Imlachs plans. Nicholl.

My advertisement. the terms not difficult. begin to approve of the measure.

News of Dr. T.

Weight removed from my mind. could I stay so till I can make my bread.

Preaching. Collection of people. Ball at the Ellens [?]

Invereschie.

Walk around the Lakes.

[1] This Andrew appears to have been a servant charged with the upkeep of Sharpey and Thomson's classroom in Surgeon's Square, Edinburgh. He is probably *not* the Andrew [Wood] who appears later in the correspondence.

[2] Possibly the David Blair Ramsay mentioned in a later letter: see note 10.1 below.

[3] Possibly William Macdonald (1791–1875), who lectured on anatomy in Edinburgh *c.* 1838.

[4] Peter D. Handyside (1808–81), private lecturer in anatomy and surgery in Edinburgh.

[5] John Russell, sixth Duke of Bedford (1766–1839), the father of Lord John Russell, who figures prominently in the later correspondence. Thomson probably owed this valuable connection with the Russell family to his father's prominence in Edinburgh Whig political circles: Lord Holland had recommended Thomson to Bedford.

[6] Henry Imlach (1815–80) graduated MD in Edinburgh in 1836; he later practised in Liverpool.

6

Remember me most kindly to Carswell and all friends.

The Doune Lynwilg (by Perth), 6th October 1836.

My dear Sharpey,

I have the offer of an opportunity to London and take advantage of it to write you a note more for the purpose of demanding your news than that I have any of interest to communicate to you.

I was obliged to you for your note before you left Edinburgh and for your arrangements about the Museum which I trust will all turn out right. I hope that the leaving the preparations will not be inconvenient to you.

You must tell me how you feel in your new situation; how your preparations for lecturing go on; what you are to lecture upon; what your University constitution is doing and in short all about yourself that you can cram into a letter; for you know that I shall feel as much interest in your welfare in your present situation as I should have done had you still been at No. 9. It is indeed a sad change in that quarter. I cannot reflect upon it without the most melancholy feelings. We go from this to Ireland and in this way avoid Edin. on our return South which I am almost not sorry for; so vexed should I have been to see Surgeons Square without my being there in my usual occupations.

My advertisement arrived here and was read aloud by the Duke in the drawing room from the Caledonian Mercury, the other night, which annoyed me excessively as the date of my beginning my Anatomical lectures was stupidly made for this in place of next winter.

I was much obliged to you for managing Macdonald. I suppose if he begins this winter he will be absent from Edinb. or off the field before the next.

You will be surprised to hear that I have been very happy here; indeed it makes me wonder myself. I daresay a great deal depends on my natural buoyancy of spirits which from my improving health has regained the ascendancy. I am as fat as a porpoise and stronger than I can recollect to have been.

The field sports interfere dreadfully with the paper on Generation. I fear I must be greatly behind, but I have had no note of hurrying from Dr. Todd.[1] I wish you could find out & let me know if I am in a scrape. About half is written, and perhaps the most difficult part viz the introductory and general part. The rest I have still to do concerns the functions of the male and female sexual organs in Man – or in other words, Conception and fecundation. It will not be good.

Tell me in your letter what I should do to prepare for my Anatomy Courses. I fancy your mind will be in this subject at present.

We shall return from Ireland about January and I still hope to be some time with you in Spring.

Do write me soon and if you are so much occupied as not to be able to answer all my queries, write me a short note telling me how you are doing.

Believe me my dear Sharpey

Ever affectionately yours

Allen Thomson

---

[1] Robert Bentley Todd (1809–60), Professor of Physiology and General and Morbid Anatomy at King's College, London; editor of the *Cyclopaedia*, op. cit., note 2.14 above.

7

25 Dover Street, London
2nd December 1836

My dear Thomson,

I shall not waste space and time in vain apologies for my long silence, but proceed at once to say that things are going on with me here very comfortably. The class room is full every day, I like my students and I think they like me. They, I mean the majority of them, are really a hard working set of fellows, and I really begin to think favourably of the prize system at least when conjoined with that of giving honours as adopted here which really seems to have a pretty general effect in a class and raises the standard degree of diligence.

I have got through general anatomy, to which I devoted a pretty long allowance of the course, and am so far with the particular functions. I describe carefully the organs concerned, especially as regards their internal structure, but I see clearly that my course will be very free of mere descriptive anatomy. I should rather say external and topographical anatomy, I shall have nothing to do with the bones, the muscles of the limbs, nor their vessels and nerves. The course will be one of physical and physiological Anatomy – to compound small things with great on the plan of the Elementa of Haller.

I miss Andrew sadly, I can get nothing on a few hours notice, and the distance to the rabbit market, the slaughter house &c. are very distressing — I think I must send to Edinburgh for frogs, I had but one decent sized fellow which a pupil brought, and after showing the circulation &c., I used him for a filtration of the blood. [F]or you must know I took in the blood in the last article in my general anatomy, reserving its charges from art[erial]. to ven[ous]. &c. till I got to respiration. In this way the processes of digestion and sanguinification will be easier followed. Our Museum is a showy one but though made principally by "*My uncle*"[1] or under his superintendence, I can assure you it is anything but a good working one. We have a man (a capital fellow) busy supplying good things for it; he has lately taken to inject lymphatics and has been most successful (with my old apparatus)[.] [W]e have now preparations of the absorbents[,] of the skin[,] subserous tissue, glans penis &c. &c. and some spendid things from the intestines of the Ostrich; it was this began it, we happened to get an Ostrich, and knowing that the lacteals could be filled on the intestine as in reptiles, I injected a bit very nicely but our friend's soon surpassed mine and has taken such delight in it that he has gone on ever since notwithstanding my double warning first against mercurial enthusiasm and secondly against the *monomania* which as you well know is so apt to seize on those engaged in hunting out lymphatics.

You must come to us and you shall have a quiet place and fresh viscera, penises, mammae, bits of skin &c. &c. from the hospital, from which you may make something for your campaign next year.

But oh I find such a great quarry of a place by no means so convenient for the daily preparation of lectures, and it will take some time before I get all well organized.

I have a young man engaged to draw for me, he makes the sketches on the board, copies them afterwards into a book, and enlarges them and renders them permanent at his leisure. I have introduced the oil painting but we are at a loss about some points on

8

which I must beg the favour of you to write me *most particularly* and in any other you may think useful. How is the canvas prepared? [I]s it done over with size or anything to prevent sinking before the ground is put on? How is the ground or indeed the colours generally rendered dull (not to shine)? How are they made to dry speedily? How long should a man take to paint such a thing say as the large oil painting of the Eye or Ear? I give 5 pounds a month and furnish materials, and the youth works from 10 to 4. From 12 to 2. I generally have him sketching on the *board* (I wish I could say *slate*) which he certainly does capitally; but we are somewhat at a loss in what colours to represent objects in the pictures. I would say to you that whatever you paint make it *large*, it is a great relief to the eye in a large room and you will yet have to lecture in some other place than Surgeon's Square. Talking of this it seems curious but it is the fact in my case that I feel less diffidence before a *large audience* now that I have got over the first trepidation, explain it as you best can.

I am living with Willis[2] who has taken a house in Dover Street; he has been *very ill* poor fellow, but is now getting sound. I mean to take a home next Summer in some quiet place.

Now that I have said enough of myself let us talk of you. First as to your Article. — The coming No. of the Cyclopa. has been delayed by Partridges[3] *Ear* and T. Bells "*Edentata*".[4] I suppose the latter are waiting for a full set of *Mineral succedaneum* from the hands of the author, but however this may be, poor Sherwood has been disappointed and Willis tells me and authorize me to inform you that there is no hurry for your "Generation" we will let you know when it is wanted; nevertheless I will forward your letter to Todd.

Keep your paper your own time and revise it to your content. I have ordered Flourens[5] which will be *very soon* procured and dispatched by Belfast. A second edition of Burdach's first volm.[6] is out, I have not seen it, but Grant[7] reports on it not favourably.

I am joining the Med[ico]. Chir[urgical]. Society chiefly for the library though it is by no means a good one. Oh how I long for the flesh pots of Egypt in the shape of the Medical Society & College Library of Edinb. Here one can't step in after lectures and ask Lewis[8] for the sight of a book. So much is this the case that I am ruining myself buying books, and mean to have a complete set of the French and German periodicals which are more immediately connected with my subject, such as the Annales and Memoires des Musee and the Annales des Sciences Naturelles *which I have got already* and Meckel's Archiv [für Anatomie und Physiologie].

In reference to your materiaux for next session, I would advise you to get Bourgery's[9] plates for the bones, joints[,] muscles and vessels, and paint enlarged copies in oil for all these departments as well as some good ones of the other parts of Anatomy which of course I need not remind you of. This plan will illuminate a dry, at least a tedious part of your course. Quain has got such things and finds them of immense use, and you will have your slate *free* for plans &c. that occur to you at the time. But I would cancel three fourths of the small branches of vessels given by Bourgery.

A youth[10] here who has been in Germany is translating Müller's Physiologie,[11] he knows the language and makes a *very literal version* but I fear it will require much

9

draping to make it presentible [*sic*] to English readers. I have nothing to do with it further than that the man was introduced to me as an old pupil of our school and a friend of Quain's. He is to put in a few cuts and note here and there. A thing occurs to me at present. If you would publish the article ovum as a separate book, I would engage to sell at least 100 copies a year among my pupils. Could you not arrange with Sherwood and Todd, but indeed you have nothing to do with them[.] [P]repare and print it as your own[;] there is no harm in this. Or will your new employment allow you time for a system of Physiology? I feel the greatest difficult in recommending a text book for Physiology. The only extent to which I at present see any prospect of myself being able to supply the deficiency is in so far as concerns General and Physiological Anatomy — Müller if *well* done into English shall be my textbook for Physiology next year. A new Edit. of Bostock[12] in one volume is just out, loaded with literature and therefore of little use to the student as a guide to the *best* sources.

I fear that in writing to Ireland I am half Irish myself, at least if an unconnected prattle is any indication – But before ending I must not forget your question as to our accounts.

The fees for last summer were:

£48.19	
£17.18	−£17.18
3 | £31.01	
£10.07	

Andrews wages . . . . £8.10 Which is your share of the summer fees
Advertising . . . . £ .14
Current expenses }
Coals etc £4. 4
Parts of body and
Logs £2.10
Spirits (say?) £2.00

I paid P. Forbes an account in August which I have mislaid but I guess the spirits used in dissecting room at somewhere about Two pounds.

I am in no hurry for payment if it is not perfectly and entirely convenient, for I may now finger a part of my London fees.

Let me know how long you will be in Ireland and whether there is any other book we can send you a look of. I wish I had had the benefit of your assistance with the drawings for Echinodermata I fear from the proof that they will turn out queer things. The proofs on india paper are very good, but the ordinary printers bungle them abominably.

I wish you might suggest some good things for the ovum and development which I might get enlarged for my lectures, think of this like a good fellow, and with the sincerest wishes for your welfare in general and your success as a teacher in particular I ever am My dear Thomson

Very affectionately yours

W Sharpey

Write soon again

[1] Apparently a reference to Charles Bell (1774–1842), the first Professor of Physiology and Clinical Surgery at the University of London.

[2] Presumably Robert Willis (1799–1878), Librarian of the Royal College of Surgeons of London.

[3] Richard Partridge (1805–73), Professor of Descriptive and Surgical Anatomy at King's College, London. He contributed the article 'Face', not 'Ear', to Todd's *Cyclopaedia*, op. cit., note 2.14 above.

[4] Thomas Bell (1792–1880), Dental Surgeon at Guy's Hospital and Professor of Zoology at King's College, London.

[5] I.e., a work by the French physiologist Marie-Jean-Pierre Flourens (1794–1867).

[6] Karl Friedrich Burdach, *Die Physiologie als Erfahrungswissenschaft*, 6 vols., Leipzig, Leopold Vos, 1832–40.

[7] Robert Edmond Grant (1793–1874), Professor of Comparative Anatomy and Zoology at University College London.

[8] Presumably, Douglas Lewis, Assistant Librarian to the Royal Medical Society, Edinburgh.

[9] Jean-Marie Bourgery, *Traité complet de l'anatomie de l'homme comprenant la médecine opérative*, 7 vols., Paris, C. A. Delauney, 1831–54.

[10] I.e., William Baly (1814–61), formerly a medical student at University College.

[11] Johannes Müller, *Handbuch der Physiologie des Menschen für Vorlesungen*, 2 vols., Coblenz, J. Hölscher, 1835–40.

[12] John Bostock, *An elementary system of physiology*, 3rd ed., London, Baldwin and Craddock, 1836.

8

[Notes written on the wrapper of the letter above.]

Answered 24th December.

glad to hear from your own mouth the account of yr. success. gives me a good opinion of yr. students embarrassment. I know what it is. popular prize system charming subject.

Difficulties of preparing for lecture. Sympathies with frogs from W. Thomson.

Lymphatic preparations – apparatus – India rubber

Drawings – prepared canvas too expensive – patent Canvas preparation of – paint – sugar of lead. using turpentine & Japan – don't skin or dry quick & wash out easily – your terms cheap. McCartney got /./ a week, but much more.

Drawings for me – copies of Quain's and yours – suggest for the ovum – there are none.

Could not get physiol. out satisfactorily. Article ovum consider must laid on shelf.

Flourens by Dublin

Willis paper formation process. Epigenesis. Evolution. bring it with me.

9

25 Dover Street

22nd March 1837

My dear Thomson,

Since you left us nothing has occurred worth noting you except that the New University have had several meetings and have at last decided on the salary of their Registrar or Secretary. Warburton and Somerville's other friends wished to fix it disgracefully low and thus keep worthier men out of the field – they proposed £200 a year to begin with, a sum for which you can scarcely procure a common clerk in London; but they have been victoriously beaten and £1000 a year is fixed on, if the Government will give it. If Forbes[1] gets the place he must I suspect give up his connection with or rather superintendence of the Journal – at least nominally – I do not know his sentiments.

I have got Coste[2] for you and will send it. Pross's prices are as follows. Stand £14. Compound body and *2* eye pieces £1"2. Micrometer Eye piece 18/6. – 1/2 inch focus achromatic object glass £3"7"6. 1/4" ditto £4"7"6 – Condensing lens on a stand £1"1.

I tried Humboldt and Müller's experiment on the frogs muscles[3] and succeeded perfectly – it was a large female full of eggs one of those you gave me.

Jones[4] (formerly in Edinb) has been on his way to Germany, he has been continuing his observations on the ovum and his paper[5] to the R[oyal] Society with recent additions has now, I think, a fair chance of being published in the Transactions. Martin Barry (Monte B[l]anc) has been lecturing at the Argyle Square School (to Reid's pupils I presume) – on Development, he has published a queer paper in Jameson's Journal[6] which young Macaulay (who is here studying at the British Museum) showed me the other day, I have not read it, but Macaulay tells me he can meet with nobody who understands it. Poor Martinus climberus, I suppose if Reid gives up he will become a Lecturer.

A lot of near a dozen candidates have applied for the Chemistry chair. Fyfe[7] was here and anxiously enquired for you. — Johnstone[8] has also paid us a visit. [H]e is well spoken of in point of accuracy as an experimental enquirer, acquaintance with the literature of his subject and as a lecturer. — I wish you could obtain *for me* some information respecting Graham's[9] qualifications as a teacher, his reputation as an original discoverer and scientific chemist of course I know.

I have got Valentin's book on Development[10] – a great part of it seems to be compiled: A curious book by Schultze on the blood and another by Nape [?] on the same subject – were sent me to look at by Dr. Forbes – would you like to review them? – (you will be paid).

If so let me know when you write which in that case must be as soon as you can make it convenient and I shall let the Dr know.

I have no more gossip – say when you will be with us again, and for my sake make arrangements to be free by the middle of September at the very farthest.

In the mean time believe me ever

My dear Thomson

very affectionately yours

W. Sharpey

---

[1] John Forbes (1787–1861), editor of the *British and Foreign Medical Review*.

[2] Perhaps Jean Jacques Marie Cyprien Victor Coste, *Embryogénie comparée. Cours sur le développement de l'homme et des animaux, fait au Muséum d'histoire naturelle de Paris*, 2 vols., Paris, A. Costes, 1837.

[3] Friedrich Wilhelm Heinrich Alexander von Humboldt (1769–1859) German naturalist; Johannes Peter Müller (1801–58) German physiologist. Sharpey refers to early experiments in electro-physiology conducted by both Humboldt and Müller designed to determine the conditions under which muscular contraction occurs.

[4] Thomas Wharton Jones (1808–91), at one time assistant to Robert Knox's anatomical class in Edinburgh. After returning from a visit to the Continent, he practised as an oculist in London while pursuing researches in anatomy and physiology.

[5] Thomas Wharton Jones, 'On the first changes in the ova of the mammifera in consequence of impregnation, and on the mode of origin of the chorion', *Phil. Trans. R. Soc.*, 1837: 339–46.

[6] I.e., the *Edinburgh Philosophical Journal*, edited by David Brewster and Robert Jameson. On Barry see note 24.3, below.

[7] Presumably Andrew Fyfe (d. 1861), a private lecturer in chemistry at Edinburgh in the 1830s and 1840s, who became Professor of Chemistry at Marischal College, Aberdeen in 1860.

[8] Possibly James Johnstone (1806–69), an Edinburgh student who became Professor of Materia Medica at Queen's College, Birmingham in 1841.

[9] Thomas Graham (1805–69), Professor of Chemistry at the Andersonian College, Glasgow; he assumed the same position at University College London in 1837.

[10] Gabriel Gustav Valentin, *Handbuch der Entwickelungsgeschichte des Menschen mit vergleichender Rücksicht der Entwickelung der Säugethiere und Vögel*, Berlin, A. Rücker, 1835.

10

(17th May 1837?
25 Dover Street
Dr. Willis's? [this parenthesis is in Thomson's hand])

My dear Thomson,

I have sent you all I can find likely to suit your purpose – after all it is but rubbish with one or two exceptions.

I have sketched out a plan for you such as I pursued and have timed the first half – which you may rely upon with tolerable confidence — But I would rather counsel you to take a little time from the first half (or at least the subjects I have included in it) and add to the second —

Serous and mucous membranes		
Glands. stomach Intestines & Digestion & lacteals		7
Heart & circulation		3
Arteries		13
Lungs, Voice glottis. &c.		5
Liver, pancreas, spleen, kidney &c.		5
Bladder, genital organs		19
Male. and perineum and		13
lithotomy	4	7
		3
Nervous system	19	5
Senses, nerves &c.		5
		4
		—
		56

Female organs of generation. Development of foetus &c.

I think David Ramsay[1] took such notes as would at least be a record of the subjects lectured on with the order and time. I will enquire in Edinburgh.

13

I start tomorrow evening – & am therefore in a bustle –

With kind regards

very affectionately yours

W Sharpey

17th May London

25 Dover Street

[1] A David Blair Ramsay of Forfarshire matriculated as a medical student in 1835 and attended classes from 1835 to 1838. He does not appear to have graduated.

11

University College, 8th Dec[r] 1837

My dear Thomson,

I take the opportunity of Dr Veltan's of Bonn going to Paris to introduce him to you and make him the Bearer of this small epistle.

Miss Thomson your sister the other day on leaving Mr Mylne's to visit Miss Baillie at Hampstead sprained her weak foot, with which she has been confined two or three days at Mr Mylnes. Dr Tweedie[1] saw her & meant to take Mr Travers[2] with him in order that the Edinb. folks might keep themselves easy as to the treatment.

I saw the foot yesterday evening and it does not seem to me very seriously hurt considering the nature of the accident. I have no doubt all will go right *tho'* as you know, a sprained ankle is a tedious affair —

The N^o of the Cyclopaedia with your paper has not yet appeared but we shall convey you a copy as soon as it can be obtained thro' Baillière if no other opportunity presents itself.

Simpson[3] I hear is giving great satisfaction — The N^o of medical Students generally in Edin[b] is smaller though the diff[ce] from last year is not very considerable.

In Glasgow *where they have no Pathology* the falling off is quite fearful. Dr. Thomson[4] the Chemist ascribes it to the state of Anatomical Instruction there (this entre nous).

We are going on favourably. I have continued to show a few experiments (more on *dead* than on *living* animals however) this year which I did not think of last year – and my attendance keeps up wonderfully – It astonishes me how people can attend lectures. I never could, at least except Murray[5] the Chemist and your father's.[6] I never attended any throughout with regularity and interest. There is no doubt a great charm in hearing interesting facts for the first time – and this accounts for chemistry and Physiology. — believe me

with sincere regard

My dear Thomson

Yours affectionately

W Sharpey

[1] Alexander Tweedie (1794–1884) trained in Edinburgh as a surgeon. He came to London in 1820 and became Physician to the London Fever Hospital and to the Foundlings Hospital. He was made a Fellow of the Royal Society in 1838.

[2] Benjamin Travers (1783–1858), Surgeon to St Thomas's Hospital.

[3] James Young Simpson (1811–70), in 1837 selected as interim lecturer in Pathology at Edinburgh to assist John Thomson. Simpson was appointed to the Edinburgh Midwifery Chair in 1839.

[4] Thomas Thomson (1773–1852), Regius Professor of Chemistry at Glasgow from 1818 to 1846.

[5] Presumably John Murray (d. 1820), a lecturer in natural philosophy, chemistry, pharmacy, and materia medica in Edinburgh.

[6] John Thomson (1765–1846) successively held the Chairs of Military Surgery and Pathology at the University of Edinburgh.

12

London, 68 Torrington Square, 30 April 1838

My dear Thomson,

The date of this letter will apprize you of my delinquency of the full extent of which I am perfectly sensible and under which sense I am now smarting. I know that since your sister wrote her share you must have had repeated communication from home, and in particular I feel that the event which has taken place since then may (for I have not read Miss T.'s letter) render it peculiarly and painfully unreasonable. I first heard of Poor General Millar's[1] death, by the paper, and I have not seen any of your family since. But I leave this distressing subject and to pass over all useless excuses for my delaying till the end of our session before I wrote you, I may now speak of the present as interesting us both infinitely more than the past. As to my own affairs, they move on so smoothly and I have such an even temper, that really I have nothing of interest to report. The New University[2] have not as yet announced a medical curriculum, I have seen the proposed Examination for the degree of Batchelor of Arts, it is a strange production, such a thing as you could suppose a man to jot down roughly after first thinking of it, not that it is short, but so unequal, here details of particulars there mere heads of subjects stated, &c &c. They propose examining in chemistry, and also in Animal and Vegetable physiology, these last a mere smattering. I suspect it would have been better to have confined themselves to those kinds of disciplines which are auxiliary or subservient for further study whether professional or general, such as languages classical or modern and mathematics, natural philosophy, mental philosophy – and perhaps chemistry and left physiology to the discretion of the candidate as taste might lead him afterwards for what is required is barely equal to the level of the Bridgewater treatises,[3] and it may perhaps interfere with the *thorough* study of branches which require more severe application. You of course have heard how the election of Registrar went, at first between Rothman and Damill, the former being ultimately chosen; it was as well that Dr Willm did not come up to town on the matter, it turns out it would have been useless trouble and expense. I still think he may get the Pathology chair in Edinb. and I have a little plan in my head which I cannot think is altogether chimerical, I cannot explain it to you now.

Your generation paper gives great satisfaction, but I fear the Cyclopaedia has got grounded for a time, I suspect Sherwoods house of their failure [they] had coquetted too far with the offers made to them by other Booksellers for the Cyclopaedia and now they find themselves in the lurch, still it is a good property and of course will not in the end be allowed to drop. I think an alliance with Simpson and Reid[4] would turn out well they are both capital men. Reid has given a *first rate experimental paper* on the nerves of doubtful function[5] which you may have heard of. But my dear fellow, write me immediately to say that you are all activity and looking forward to the winter campaign, for I have seen several pupils from Edinb. (Geor. Newbiggin, Spittell – Johnny Wood[6] &c.) who ask me if Dr A. T. is to lecture in Edinb. next winter, I of course tell them – he certainly is, but I wish I could speak of the matter more closely and definitely. I am vexed that you should lose any advantage by being out of sight. Mr Wood has been here, he told me Andrew[7] was to lecture on pr. of physic – I suspect he is looking to an alliance with Peter *Simple* [i.e. Handyside], and probably because he fears you may not be able to fulfil your intentions. I have taken every care of your embryonic Prepar[ns] which are at present in my keeping, but the *double goose* is in Edinburgh. I have offered a prize in Practical Physiology and chosen for the subject on this occasion the embryology of or rather the ovology of the rabbit, from the time the ovum leaves the ovary till the 10th day – as regards the embryo and the 15th as regards the rest of the ovum. Now is the best time for laying aside for you anything in the shape of preparations — You can have what you like by paying for the *interment*. Sections of *heads*, *pelvises*, and their contents, ligaments &c. Prepared muscles &c. The only difficulty is the expense of spirits, but come yourself, and then get down to Edinb. a considerable time before the winter, I regard that as of great consequence. I have just been interrupted in writing by a call from a most excellent person – a Mr Edward Hobson[8] from Australia who has been studying here some time and has really *accomplished* himself in Zoology, Comparative Anatomy & Geology not neglecting the rest of this profession. He has gained Grants Gold Medal – and has brought various things for Owen.[9] He is to send me a collection of gravid uteri of Australian Animals in Spirits – can you suggest anything else. (he goes in 2 months) I trust Lord Glenelg[10] will do something for him[.] [H]e wishes the place of assistant surgeon to the Government Hospital at Sidney or in Hobart town in both of which there are vacancies – a place worth about £180 pounds per annum, and in which he has served already as a sub assistant. I trust the government will put him in a place where he may have the means of following out his favourite pursuits, it would be of the greatest benefit to science. Owen and Coste as you will see have been fighting about the honour of *discovering* the allantois of the *Kangaroo*.[11] To me the thing appears highly ludicrous, there might have been some interest in the dispute had they shown that the Kangaroo had no Allantois. I suppose you have *discovered* that an English Duke may survive in Nice as well as other people. But the last thing I say to you is come away as soon as your engagements will permit, and take up your position.

In the mean time believe me your affectionate friend
W Sharpey

[1] I.e., William Millar, soldier in the Napoleonic wars, who committed suicide in 1838. He was the son of John Millar (1735–1801), Professor of Law at Glasgow, who was related to William Cullen on his mother's side. One of John Millar's daughters was John Thomson's second wife and Allen Thomson's mother. William Millar was, therefore, Allen Thomson's uncle.

[2] I.e., the new University of London.

[3] The Bridgewater Treatises were a series of volumes dealing with questions in natural theology which appeared in the first half of the nineteenth century.

[4] John Reid (1809–49), at this time lecturing in the Edinburgh extramural school. It may be that there was a proposal for Thomson, Reid, and J. Y. Simpson to go into partnership; if so, nothing seems to have come of the suggestion. See also letter 15.

[5] Presumably, 'An experimental investigation into the glossopharygeal, pneumogastric, and spinal accessory nerve', *Rep. Br. Ass. Advmt Sci.*, 1837, **ii**: 109–12.

[6] Presumably George Stewart Newbiggin (d. 1840), President of the Royal Medical Society in Edinburgh in 1837. Robert Spittal (1804–52) was an early exponent of mediate auscultation in Britain: he lectured on "medical acoustics" in Edinburgh in 1838. John Gillespie Wood (1816–73) graduated MD in Edinburgh in 1837; he later became a military surgeon.

[7] I.e., Andrew Wood (1811–81), President of the Royal Medical Society in 1830 and President of the Royal College of Surgeons of Edinburgh in 1855.

[8] Edward Charles Hobson (1814–48) studied medicine first in Hobart then in London. He left England in 1839 to practise medicine in Hobart, where he also continued his researches as a naturalist.

[9] Richard Owen (1804–92), Professor of Comparative Anatomy and Physiology at the Royal College of Surgeons, London.

[10] Charles, Baron Glenelg (1778–1866), Secretary of State for the Colonies.

[11] See Richard Owen, 'Remarks on the physiology of the Marsupilia, being a reply to the communication addressed by M. COSTE to the French Academy of Sciences, entitled, "Mémoire en réponse à la lettre de M. R. Owen" ', *Mag. nat. Hist.*, 1838, **2**: 407–12.

13

24th November, 1838
68 Torrington Square
My dear Thomson,

I was much gratified with your letter — What an happiness you have, over me, that you have some one to encourage you through your anxieties and struggles and make your fireside doubly happy when the week's labours are ended. You tell me to do the same, but I am now so little in the way of looking out that I despair of emancipating myself from the regular London University College batchelor stupid sort of life I have got into; I feel therefore gratified when a friend like you in a spirit of commiseration offer a hint for my guidance – I suspect however I am too old to be a suitable companion to your young friends.

I am of course delighted to hear of your class — I wished to keep down your expectations to *a dozen* though I believed you might have *twenty*, but I assure you I never expected you would have more to begin with. I therefore regard your success as *signal* the more so as the class is nearly new, indeed entirely so, there being none of my students but Mr. Hindle[1] who perhaps you have this term discovered is a very diligent and attentive person. The first year you will recollect I had 21, I think.

Now that all uncertainty and anxiety is removed as to the encouragement you might meet with, I will take the liberty of giving you a hint. You remember a tall gawky looking lad Davidson[2] (from Arbroath) who attended your lectures as well as mine – he was here lately – much improved – and I found on conversing a little with him that his remarks on the Edin. lecturers were really very *shrewd*, and from what I knew they seemed to me to represent pretty much the opinions of the average of country students who resort to Edinburgh, when their notions have been somewhat corrected by experience. Nay I must add that you would have been astonished to hear so much sound sense come out of Mr Davidson's mouth — Now he told me that your lectures (I mean your former lectures) were very excellent as regards the information they contained, but he said your delivery *wanted force*. I have no doubt that three years more added to your life and the experience of the world as you have enjoyed it, must have supplied in a great measure what was frankly deficient; but notwithstanding all this, take the hint which I now offer you. I have heard nothing of your style since your reappearance, but within the limits of good taste you cannot be too forcible or impressive in manner and delivery. With great respect for Mrs. Siddons[3] I suspect she devoted her efforts rather to give clearness and elegance to her pupils' style of delivery — *Jones*[4] here used to abuse me if I was not forcible — he used to say "Oh! that is all nice and clear & so on, but you must remember that you must calculate on having to deal with careless auditors as well as attentive, and you must endeavour to *command* attention!!['] You remember Charlie Bell's remark that your manner was too *confidential*. Davidson added of his own accord what was very agreeable to hear, that all my students were delighted by your superintendence in the dissecting room; this is a *Hauptsache*. Let your manner be cheerful, good natured, never ironical, and in all important things serious and impressive. Don't get impatient with stupid lads, many a boy who annoys you by his apparent stupidity this year may delight you by his proficiency the next. Keep all the lounging and gossiping friends out of the Dissecting Room. And last of all dont be too anxious and distress yourself — *you are assured of success*, take care of your health.

When I look back I freely feel the truth of *Andrew's* remark, and I am conscious that I often vented my feelings of impatience or disappointment on poor Andrew. You may tell him that I was sorry for it at the time and that I have not ceased to feel compunction for it yet.

We tried an experiment here with acetate of albumin as an antiseptic, injecting 2 *bodies* with it, but the result was not very satisfactory, the muscles were rendered friable and much discoloured — If arsenic were quite safe for the dissector it would answer admirably — We had a body injected with a solution of oxide of arsenic in water (mere water) prepared by long boiling, and the antiseptic effect *was complete*. The gentlemen who dissected it made no complaint, but I find doubts are entertained how far it is quite safe — For your class subjects & many of your larger wet preparations in tubs, I am convinced it would answer admirably. [T]here is no deposit from it with the albuminous matter of the body as with corrosive sublimate, the knives are not stained, nor is the colour of the textures materially impaired, not near so much as by spirits. I will add the receipt when I return from the college. We have not finished the directions for dissection but they are not forgotten.

Would you make Andrew get for me a lot of sheep's uteri and can some of your young men search them for foetuses – I mean of course early ones – We have got difficulty here and I shall be in want of some — Talking of sheep I rather anticipated that the Apostle's[5] class would come down a little this winter, the *chosen few* will be still fewer and more scriptural next year. I am much amused by "*Dr Handyside's* case of suicide"[.][6] "Peter Martyr" or "Peter Poundtext" – I never had any grudge at him but I certainly do not disguise my dissatisfaction at some of the Balsam singers who took up his cause when he first started. My class here is quite big enough to satisfy me. I have signed my 360th ticket and we generally have a sprinkling more at the new year or before the end of the session. I was in alarm because first 50 additional seats were added last winter and this summer in consequence of an additional alteration 40 more were obtained though the addition was not courted by me. I feared naturally that with 90 new seats the room would look scarcely filled but it looks very respectable notwithstanding.

Say to Mr Wood[7] that I am obliged to him for his attention to the *tripartite treaty*,[8] or the result at least of the triple alliance. On the whole it is good though in one or two minor points I would differ — I would be against holding out *repeated attendance on lectures* as one of the grounds of superiority in the education of the M.D. Extended education if you like, but I would not proclaim that frequent *repetition* of the *same course*, unless it be a clinical or a manipulating one, is likely to make one man much wiser that another who has diligently attended once.

I am against *exacting* more than one course on any subject — Supposing however they do why lend themselves to the deception of substituting military surgery for the ordinary courses. Mil. Surg. either is the same as common surgery or it is different[.] [I]f the same why a separate course? if different why substitute it — It is a class which *civil* licensing boards should have nothing to do with. The Medical Boards of the Army or Navy may demand it with propriety or rather a course of lectures on the duties peculiar to Military & Naval medical officers – but for others to admit it as equivalent to a second course of surgery (unless in my view of leaving a second attendance entirely to the judgement of the student or his advisers) is ridiculous. The reason of the proceeding is however quite plain to me – it began by adjustments, arrangements, bargainings for support &c &c in the College of Surgeons, and having adopted it then it must now be carried through. The Three Bodies should have availed themselves of their new arrangement to throw it out. I dont blame Sir George,[9] who in arranging the University Curriculum behaved very disinterestedly in regard to his own course, but the over complaisance of his friends — I must add however that it gives me sincere pleasure nay more than I can express to remark the *spirit* in which the propositions have been conceived. So fair so free from selfishness, the *act* so conformable to the proposed intention I mean the improvement of education — What a contrast to the spirit which motivates the majority of the members of the old medical corporations in this quarter and in Ireland. The Edin. College of Surgeons has shown itself worthy of its high reputation for integrity and enlightened zeal for educational improvement.

Having spread over nearly another sheet of paper I conclude by subjoining the receipt I have alluded to – now with best regards to your wife and to Dr and Mrs W*m* in whose anxiety respecting their little girl I sincerely sympathize, though I trust that by this time her recovery is advancing – I ever am My dear Thomson

Your sincere friend

W Sharpey

Take Half a pound avoirdupois of white arsenic grind it and boil it with five pounds and a half of water for two or three hours, adding to make up for evaporation if necessary.

Inject the clear liquor by the Aorta –

Receipt for Cold Injection

Take

Dryers (a preparation of white lead and oil known by that name in the paint shops) three quarters of a pound

Red Lead half a pound

Turpentine Varnish a pint (English) and a half

Boiled oil about a Pint

Grind the red lead on a stone with a little of the oil, not so much as to make the mass very thin otherwise the grinding is difficult.

Add the turpentine varnish to the ground red lead on the stone and mix.

Mix the dryers with enough of the oil to reduce it to the consistence of cream. Add it to the mixture of the red lead and varnish and stir all together with a knife or stick then inject immediately.

The quantity of oil to be used depends on the rapidity with which you wish the injection to harden – we generally use as much as makes the mass (when all the ingredients are mixed) about the consistence of thick cream or mustard. In this state it usually acts in a quarter of an hour & runs perfectly – if you add more oil you may delay the setting for one or two hours, but though thin it will not fail to set. For the dissecting room it needs the addition of red colour the red lead being quite sufficient.

[1] Possibly Richard Hindle, who matriculated as a medical student in Edinburgh for the 1835–6 session.
[2] Possibly Patrick Davidson (d. 1852), who graduated MD at Edinburgh in 1836.
[3] Sarah Siddons (1755–1831), actress.
[4] Presumably Thomas Wharton Jones: see note 9.4 above.
[5] I.e. Alexander Monro III: see note 2.12 above.
[6] P. D. Handyside, 'Account of a remarkable case of suicide, with observations on the fatal issue of the rapid introduction of air in large quantity into the circulation during surgical operations', *Edinb. med. surg. j.*, 1838, **49**: 209–21.
[7] Alexander Wood (1817–84), Edinburgh physician and lecturer on Practice of Physic in the extramural school.
[8] These remarks seem to refer to a joint statement on the education of medical practitioners issued by the Royal Colleges and Medical Faculty in Edinburgh in October 1838. See: 'Propositions relative to the education and privileges of graduates in medicine and medical practitioners, agreed on by the medical and surgical professors in the University, the Royal College of Physicians, and the Royal College of Surgeons of Edinburgh', *Edinb. med. surg. J.*, 1839, **51**: 262–7.
[9] George Ballingall (1780–1855), Professor of Military Surgery at Edinburgh University.

14

London, 16th Feb. 1839

My dear Thomson,

I send you down a package the chief part of which is for your Brother William and one or two little things of your own.

Your letters are always welcome & and the larger the better — I have before me the one of the 23^d Dec. to which I have long been minded to write an immediate reply but various occurrences have withdrawn me from it. The egotistical part of your letter as you express it is precisely what interests me the most – and I do assure you my dear Thomson that your great success this winter has afforded me many very happy moments when I have thought on it. My remarks in my last were well meant and I rejoice to think that as I expected they have been well received – but from all I have learnt I am inclined to think they were unnecessary tho' as perhaps serving to confirm you they might not have been superfluous.

Monat [?] who is man of judgement and taste as to manner at least whatever he may know of the matter of a lecture has given me a most favorable report of a lecture which he heard of yours – the more favorable in my opinion as it is judicious and he speaks specifically — Dont let yourself flag at the end of the session. You amuse me by your account of your *distressing mistakes.* I mean distressing to yourself – be assured they are of no consequence — But you astonish me by your diligence – in the midst of a first course with all its anxieties and needful preparations and with all the trouble and fatigue and necessary occupation of the dissecting room — I say you astonish me that in such circumstances you have been able to do what you have done Ligaments – viscera – preparations of teeth – drawings, new contrivances &c. &c. it is scarcely credible. Many thanks for the young lamb, it is small ones that I want – pray procure *several* stout bottles and pack them with wadding and spirit – not too many in each bottle – make Andrew keep my account.

Among the drawings you took with you there were some for which you or your father had a special regard some of those of Dr. Gordon's[1] and which I understood you were to retain – but so soon as you can spare the others I would be obliged to you to send them as I am hastening towards that part of my course and I fear I shall be destitute — I shall get those of *Reil's of the Brain*[2] done here anew so you need not send them, nor the base of the brain from Gall's[3] views, nor Vicq d'Azyrs[4] – indeed *none of those of the brain.*

Have you any observations on the formation of the Amnios in the human ovum? I see these Frenchmen have gone back to Pockel's view (chiefly Ibelin) – because they have seen a sac within the chorion & without any embryo – which they think must be the amnios into which no embryo has *passed.*[5] The interior sac in such abortion ova may be the enlarged vesicula umbilicilis may it not? The best of the joke is that Breschet & Serres[6] are disputing between themselves about the priority of what in truth belong to Pockel if there is any merit in it. They must be mistaken say I.

I send you down an article by Henle[7] which will interest your Father — I do not know the author of the article Magnetism in the review.[8] Baly has out another part[9] which I shall enquire after tomorrow and if possible send in the parcel – (I find it has been sent you [by?] Mrs Maclachlan)

You will see we have done with magnetism at the sacrifice of Elliotson[10] — It is a blessed thing for any body is better than he had latterly made himself — In many respects he was in my opinion all along an objectionable person – but he was a good teacher and a *man of note* – enjoying a large reputation in *the country* and as such useful. [B]ut after his absurdities of the magnetism & the scandalous proceedings carried on by him in the Hospital he could be regarded as nothing better than a broken pitcher which could no longer *hold in* or be of use to any body — You would hear of our *emetic?*[11] it was not so great as some people represent it, and the second night all was quiet – *perfectly so* and has been ever since.

What is this University Club?[12] Let me know about it because I have had a letter from Mr James Macaulay[13] advertising it, and if it is not calculated to interfere with the existing societies, if they do not intend having a library or evening meetings to discuss papers – I think it would be stiff to withold my subscription. The only think makes me doubt is that Peter Pompous [i.e., Handyside] presided over their constituent meeting, and it occurred to me that most probably the promoters of it were of the same stuff.

I had a letter some time ago from Mr Slater the sculptor wishing to know my directions respecting that famous production my *bust* which is still on his premises, 50 George Street – be so kind will you as look after it; make him pack it up and send it to my sister Mrs Colvill Arbroath, and might I beg of you to pay the expenses of packing and carriage — there is no hurry with this indeed it would be better to defer it for a couple of weeks, as I shall be writing to Arbroath before then.

With this there are some copies of an Introd. Lecture by Dr. Leighton,[14] a friend of mine – & though some of the views defended in it appear to me heterodox never the less the spirit of it and general tone is altogether commendable – especially as emanating from a *London* teacher.

23^d Feb^r

I was late with my parcel — I send your father a little work by Dr. Gluge[15] which may entertain him – I happen to have another copy which the Author sent me (from Brussels where I believe he is professor) so the Dr may add it to his library –

I return to you also some of your own books which I have kept too long. —

You will find the piece of stomach you wished for, also a very putrid bit which I put aside for you as showing the true Brunners glands of the Duodenum but I fear it has suffered so much as to be now nearly useless —

I had a letter from Syme[16] since this letter was composed, he says you have a fair class and an excellent reputation giving great satisfaction to the students. He adds that you are sure of Glasgow – that he had been *there* lately and that your appointment was considered a settled – and as far as I gather from the content a desirable thing – I know there are *loud complaints* by Jeffray's[17] colleagues at the great detriment the school is suffering by his inefficiency –

> With every good wish believe me ever my
> dear Thomson
> Your sincere friend
> W Sharpey

I send you a model in wax of a diseased bone (exfolient) we have one already in the museum & I have no use for more.

---

[1] I.e., John Gordon (1786–1820), private anatomy lecturer in Edinburgh from 1808 to 1818 and author of works on the structure of the brain.

[2] Probably a reference to: Johann Christian Reil, 'Fragmente über die Bildung des kleinen Gehirns im Menschen', *Arch. Physiol.*, 1807–8, **8**: 1–58.

[3] Franz-Joseph Gall, *Sur les fonctions du cerveau et sur celles de chacune de ses parties, avec des observations sur la possibilité de reconnaitre les instincts, les penchans, les talens, ou les dispositions morales et intellectuelles des hommes et des animaux, par la configuration de leur cerveau et du leur tête*, 6 vols., Paris, J.-B. Baillière, 1825.

[4] Félix Vicq d'Azyr, *Planches pour le traité de l'anatomie du cerveau*, new ed., Paris, Louis Duprat-Duverger, 1813.

[5] See: [?] Pockel, 'Neue Beiträge zur Entwickelungsgeschichte des menschlichen Embryo', *Isis*, 1825, cols. 1342–50; Gilbert Breschet, 'Remarques sur la communication faite par M. SERRES concernant le développement de l'amnios chez l'homme, *C.r. hebd. Séanc. Acad. Sci., Paris*, 1838, **7**: 1031–8; E.-R.-A. Serres, 'Observations sur le développement de l'amnios chez l'homme', ibid., pp. 996–1000.

[6] Gilbert Breschet (1783–1845), French anatomist; Antoine-Etienne-Reynaud-Augustin Serres (1786–1868), French comparative anatomist and embryologist.

[7] Friedrich Gustav Jacob Henle (1809–85), German anatomist and pathologist.

[8] [Anon., Review of A. Mesmer, *Mémoire sur la découverte du magnétisme animal*], *Br. for. med. Rev.*, 1839, **7**: 301–52.

[9] I.e., Baly's translation of Johannes Müller's *Elements of physiology*, op. cit., note 7.11 above, published in 2 volumes by Taylor and Walton (London) in 1843.

[10] John Elliotson (1791–1868), from 1831 Professor of the Practice of Medicine at University College London. He resigned in 1838 after differences with his colleagues over his use of mesmerism in the treatment of patients at University College Hospital.

[11] Probably a reference to the strategy employed by the Hospital's Medical Committee to force Elliotson's resignation. It ordered him to discharge Elizabeth Okey, one of his patients, who allegedly displayed the mesmeric gift of prophecy. This was coupled with a request that Elliotson cease to employ mesmerism on the wards. On receiving these instructions, Elliotson resigned.

[12] I.e., the Edinburgh University Club, founded by Edward Forbes and others in 1839.

[13] James Macaulay (1817–1902) studied medicine in Edinburgh, but later turned to literature as a career.

[14] Frederick S. Leighton, *On the objects and mutual relations of the medical sciences: an introductory address delivered at the Middlesex Hospital School of Medicine*, London, H. Renshaw, 1838.

[15] Possibly, Gottlieb Gluge, *Observationes nonnullae microscopicae fila (quae primativa dicunt) in inflammatione spectantes. Dissertatio inauguralis pathologico-anatomica*, Berlin, Nietach, 1835.

[16] James Syme (1799–1870), Professor of Clinical Surgery at Edinburgh since 1833. He became acquainted with Sharpey while both were students in Paris in the 1820s.

[17] James Jeffray (1759–1848), Professor of Anatomy at Glasgow from 1790 to his death.

15

[Thomson to Sharpey, undated fragment, presumably February 1839; see the letter above.]

me as I daresay it will be to you a sufficient proof of the erroneous nature of Pockel's view as well as of all french and english "Mechanics"[1] Coste has committed himself in this first volume beautifully on the matter & the best of it is I shewed him the error on his own specimens, after which he spoke so that you would have thought mine had been his opinion all along and that he had never participated in that of Pockel. They are a set of villains.

I think I told you that I found the spinal cord open in Reids[2] specimens w^h appeared to be 14 or 15 days old — Cumins[3] specimen presented only the primitive folds and groove —

I am glad that Elliotsons row is over you must have been in a state of prodigious excitement to utter that defiance of all the dastardly cowards sneaking in corners which the newspapers give. Who is likely to succeed?

Simpson has met with more success than ever attended any midwifery teacher or practitioner in Edinb. but he exerts himself too much & has too many irons in the fire. Inoculating chancres here, proving the truth of magnetism, then disproving it in another quarter &c &c &c.

We are going to have a monthly journal here in which a dozen of us have embarked for next session. Medical records in the line of the Lancet but all scientific. J. A.

Robertson,[4] Simpson, Henderson,[5] Balfour,[6] Duncan Seller[7] & myself with others. John Reid and I having the Anatomy and Physiology department.

By the bye the Lizars are to leave Argyle Square next season and go down to No. 1 in Surgeons Square which has been bought by an Aunt & is to be remodelled. This of course throws the great school into consternation. Robertson I suppose will do the Surgery there but they are at a loss for Anatomy. He has been trying I think to persuade Reid to start there in Anatomy but Reid I suspect declined. Reid would fain join with me and in fact the principal obstruction is the regulation of the College about two subjects by the same teacher. I have not seen him to speak to him, but I think we shall make some sort of arrangement together which would strengthen both of us considerably.

The University Club I know little about. They are only I believe to have journals & newspapers & a reading room. I do not hear of memoirs to be read or discussions. I think you may subscribe. I have never been applied to.

I should like to consult you about a great many Anatomical points as I used to do when we were together, but they are so numerous I fear I must leave all this till I have an opportunity of conversing with you. There is one however I must ask you to

---

[1] See note 9.2 above. These embryologists were "mechanics" because they maintained that the amnion was produced by a process of endosmosis.

[2] Possibly Henry Reid (d. 1868), President of the Royal Medical Society of Edinburgh in 1842.

[3] Possibly Joseph Edward Cummins, who graduated MD in Edinburgh in 1837.

[4] John Argyll Robertson (1800–55), private lecturer on surgery in Edinburgh.

[5] William Henderson (1810–72), Physician to the Fever Hospital and Pathologist to the Edinburgh Royal Infirmary.

[6] John Hutton Balfour (1808–84) graduated MD in Edinburgh in 1833. He founded the Botanical Society of Edinburgh in 1833.

[7] Perhaps a mistake for *William* Seller, who graduated MD Edinburgh in 1821. He was Physician to the Royal Public Dispensary and Extra-Physician to the Edinburgh Royal Infirmary.

16

London 9th March 1839

My dear Thomson

I now write you confidentially I mean only for yourself and immediate friends, because unless perchance called upon by yourself or others I dont wish to be understood to interfere in the matter to which this letter refers — The thing may after all be *superfluous* for it is many chances to one that you are far better informed than I am on the subject of *Jeffray's Chair*.[1]

I have just heard of his serious illness – from which it is said he will not recover, and I write to put you on your guard against anticipation, because I know that interest is now making for Dr. Mackenzie[2] as a successor. Application was made to Sir James Clark[3] to interest himself for Mackenzie, and I am extremely glad to say that he told the party applying, that he had reason to think the chair would likely be conferred on another person whom he considered as deserving of it, and in the circumstances he thought Mackenzie's application would be useless; moreover that seeing what his own opinion was he could not think of interfering on Dr. M's account. The person he meant as likely to get the appointment and so well qualified for it was yourself.

Now this may turn out to be "*Piper's news*" that is that you know it already, but I will put you to the expense of a postage because I think it important if in the event of Jeffray's death you should be required by Lord John[4] to *give a reference*, that you should be aware of Sir James's sentiments respecting your fitness, for he would be a very proper referee —

I hope my present arrived without accident — Pray dont forget to send me the drawings of the *Ear*.

Give my best regards to all your household especially Mrs. Allen – and I say I anticipate the pleasure of soon paying her a visit in an old fashioned domicile not very distant from a certain very shabby street in the City of Glasgow.

Let me put in a word for *Andrew*.

Wishing Jeffray a most comfortable transition to a *better place* and in the mean time your own advancement to a *very good one*. believe me my d[r] Thomson

very truly yours

W Sharpey

Dr Allen Thomson

[1] I.e., the Chair of Anatomy at Glasgow University.

[2] William MacKenzie (1791–1868), Professor of Surgery at the Andersonian College in Glasgow from 1819.

[3] James Clark (1788–1870), Physician-in-Ordinary to the Queen.

[4] Lord John Russell (1792–1878), the Home Secretary, had been a student in Edinburgh where he moved in the same Whig circles as John Thomson.

17

68 Torrington Square, London

4th Dec[r] 1839

My dear Thomson,

I was delighted to hear from you, and to find that altho' *for the present* things are not so flattering as they might have been yet you are *true to your Philosophy* and meet all in a proper spirit. Your school is not singular. There is a defalcation in London this year and we bear our share of the loss, no better I dare say from the crisis we passed through last winter by the circumstances *preceding* & *accompanying Elliotsons secession.* Diminished numbers of new *entrants* in the profession which is now beginning to tell on the number of *students*, the pecuniary pressure in trading and manufacturing districts from which we receive a large proportion of students, the advantages such shortness of money gives to Provincial schools established chiefly in those very districts and encouraged to an injudicious extent by most of the Examining Boards – all these things combine to hurt us at present but things will get better again – So you must not be disappointed. The Aberdeen folks will not persevere in passing the door of what will before next year be proved to be the better shop, at least transcendentally the best in Anatomy & Physiology, so you will have it all your own way.[1]

I think you had better follow Mr. Bannerman's[2] advice and dispose of your collection – wherever else you are likely to go there are collections already, and one thing is clear that when a public collection may be made I would advise no-one to collect one or continue collecting for himself. I restrict myself to Embryology in order to have something to show as my own, but it is folly to attempt a general anatomical collection for oneself. I should have difficulty in advising you as to the price to ask for it unless you send me a catalogue of the preparations. I should have great delight in seeing some old friends my own handiwork occupying a distinguished place.

Talking of Embryology — I don't believe Todd can be in any hurry for your article, but I am very sorry it is not ready for printing and separate publication, for I would not like the edge to be taken off it by Müller's Chapter in his Physiology – that will not be in the next part however it is not yet out in German even – the next part will contain generation and a rigmarole about the mental faculties which he had better left out as forming a separate science. But *Willis* has translated Wagner's little volume[3] and I understand that Todd having done a little of it will *father* it all, so that it will appear with the Priest's arms (Kings College) upon the title page.

As the time of post approaches I am obliged to close my letter – but now that we can communicate for 4[d] I will again soon write – indeed one of the consequences of cheap postage will be that we will write shorter letters and more of them –

With sincere regards to Mrs Thomson I am My dear Thomson

Your sincere friend

W Sharpey

---

[1] Thomson had in October 1839 been appointed Professor of Anatomy in the Marischal College, Aberdeen.

[2] Possibly John Bannerman, mediciner of King's College, Aberdeen.

[3] I.e., Robert Willis's translation of Rudolph Wagner, *Elements of physiology, for the use of students, and with particular reference to the wants of practitioners*, London, Sherwood, Gilbert, & Piper, 1841.

18

68 Torrington Square, London

28th November 1841

My dear Thomson,

I write you hastily to say how delighted I am to hear of your signal success – it is indeed *most important* for you as regards both the present & the future.

Your present has arrived safe – the Embryos are beautiful – and I beg you to accept my best thanks for the gift – and to confer an additional favor by replying to a few questions – viz. –

What is the liquid used? is it Spirit or Goadby's Saline fluid (of which I have the recipe)? for what preparations is each best suited? How do you deal with the liquid in closing your preparation? Do you smear the edges of the lid with the cement previously to putting it on, or do you put the lid on clean – dry the edges, or allow them to dry, and then apply the cement outside?

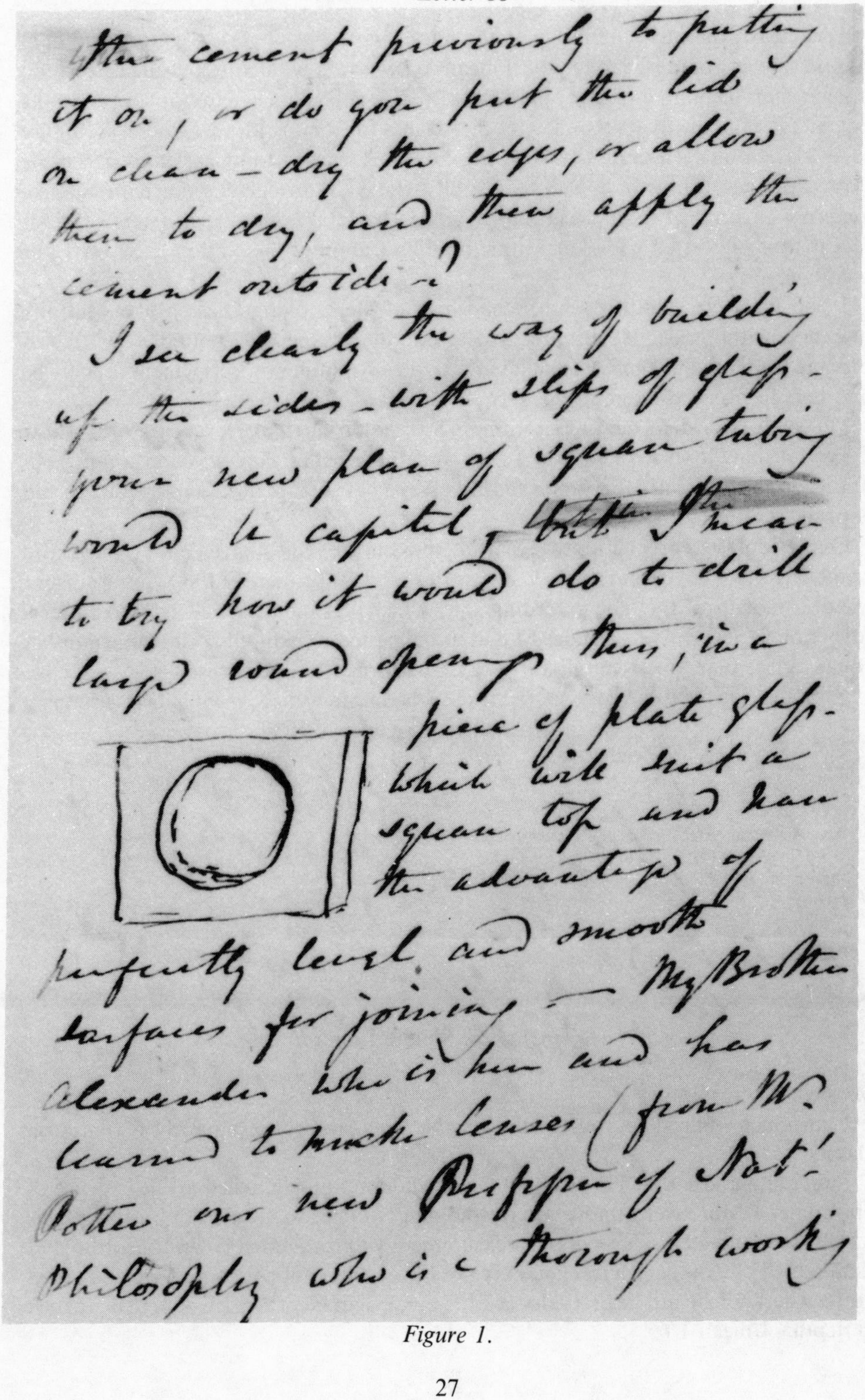

the cement previously to putting
it on, or do you put the lid
on clean — dry the edges, or allow
them to dry, and then apply the
cement outside?

I see clearly the way of building
up the sides — with slips of glass —
your new plan of square tubing
would be capital, but I mean
to try how it would do to drill
large round openings thus, in a
piece of plate glass —
which will suit a
square top and have
the advantages of
perfectly level and smooth
surfaces for joining — My Brother
Alexander who is here and has
learnt to make lenses (from Mr.
Potter our new Professor of Nat.
Philosophy who is a thorough working

Figure 1.

27

I see clearly the way of building up the sides – with slips of glass – your new plan of square tubing would be capital, but I mean to try how it would do to drill large round openings on them, in a piece of plate glass, which will suit a square top and have the advantage of perfectly level and smooth surfaces for joining [drawing here, see figure 1] — My Brother Alexander[1] who is here and has learned to make lenses (from Mr Potter[2] our new Professor of Nat*l* Philosophy who is a thorough working man) tells me he thinks he can easily bore the glass provided he has a large enough drill; it is a curious fact that when wetted with Turpentine holding Camphor in solution glass works as readily as brass.

I find Plate glass cuttings can be had here of the Looking Glass Makers, of any thickness within reasonable limits, at the rate of nine pence a pound. I will try to procure for you some *thin ground* glass for covering minuter objects which can now be had here in broad pieces and not dear.

Do you first harden your preparations in Strong spirit to make it keep its form – how do you display it – I wish we could find something better than wax for pinning out upon – for strong spirit almost always dissolves a part of it & occasions a white turbid deposit.

Our school is greatly better this year in New Entries which have kept up our entire number rather above what it was last year tho' we lose the benefit this year of most of the old students of the time preceding 1839 which was the year we began to suffer a reduction in new entries — Indeed I had not expected to maintain our whole number equal to last year, but I am glad to say it has turned out otherwise —

With regards to Mrs Allen and the rest of the family believe me my dear Thomson
Most sincerely yours
W Sharpey

[1] I.e., Alexander Arrott, Sharpey's half-brother.
[2] Richard Potter (1799–1886), Professor of Natural Philosophy and Astronomy at University College London from 1841 to 1843, when he moved to Canada.

19

68 Torrington Square London
27*th* June 1842

My dear Thomson —

I am much obliged to you for the trouble you take concerning my preparation glasses.

I am quite aware that I ordered 30 dozen and am not frightened at the expense of "bottomry". Moreover I thank you for the suggestion as to the sizes of bottom plates which I would prefer to have of the measurement you recommend – and accordingly I shall return you the sketch lest you may not have kept a note of them. I see you are in doubt whether I meant the first 4 dozen of 3/4 inch diameter to be 1/6 or 1/16 of an inch in depth – I mean 1/6.

Letter 19

I have heard from what I consider *very good authority* that arrangements are about to be made for Homes'[1] resignation and therefore I presumed it has been determined on — In such event it is clear Alison's[2] Chair will be vacant, and I am satisfied that you not only have the best claim to it but will have the best chance of it — I should think from what I have repeatedly heard expressed on former tho' not very distant occasions that your appointment would meet with approbation from all the men in the University; and were you judiciously and timeously to strike in you would I think have scarcely an opponent — The great point will be to avoid being *taken up*, as the phrase is, by a particular party, and I think from the way you stand with every body you are likely to be the general favourite and have no occasion for party support. The excitement of an Election in Edinburgh is, however, ready to give rise to indiscretion if not in the candidate at least among his friends (witness Miller's[3] friends) fit to peril the very best prospect – Midwifery, Surgery, – Pathology, Practice of Physic & Physiology! – a few more changes & the College will have got a new Stock a new lock and a new barrel. — Do not take it amiss if I entreat of you to caution your immediate or more active supporters to make no disparaging remark of other candidates – not that I have any *special* reason for doing so but being myself a looker on I take the liberty of offering this hint solely from what I have witnessed in other cases & from the *possible nature of the objections* that may perhaps be raised against some of the parties who may be your competitors – J. Macfarlane[4] "of the Bridge" called and left his card a day or two ago – I wish I could see him.

With best regards to Mrs T. & my friends at Morland – I am,

My dear Thomson
yours very sincerely
W Sharpey

After putting on the top do you let the edges dry completely before you use the cement? — I find if you wait till the surface to be covered with cement is quite dry that a portion of the fluid between the contiguous surfaces or even within the cell is apt to evaporate —

WS

[1] James Home (1758–1842), Professor of the Practice of Medicine in Edinburgh from 1821 to 1842.

[2] William Pulteney Alison (1790–1859), Professor of the Institutes of Medicine in Edinburgh since 1821; he succeeded Home in the Practice Chair.

[3] Presumably James Miller (1812–64), appointed Professor of Surgery in Edinburgh in 1842.

[4] Possibly John Macfarlane (1796–1869), a surgeon at the Glasgow Royal Infirmary. The "Bridge" may be a reference to one of a number of local villages.

20

London 10th Aug 1842

My dear Thomson

I fear you will think I have forgotten your letter, but it was only today that I made out a call on Dr. Hodgkin.[1] The good little Quaker, it seems, declined giving Davy[2] a Testim[l] for the Pathology because he had previously given a strong one in favour of Craigie[3] but offered Dr. D. his services in the event of his applying for another chair, & as the Physiol^y Chair was meant, he thinks were Davy to be a candidate & find he had given you a testimonial Dr. Davy might justly complain of there being a different rule applied to him & to you — From a letter however which Dr. H has rec^d from Dr. D. it seems unlikely that Dr. Davy will apply for the Physiology Chair and if Dr. Hodgkin is authoritatively informed of Davy's determination not to offer himself he will then give you a fresh testimonial of the present date, and express in it what he *at once and unhesitatingly* declared to me orally, namely, that he knew no one in the three kingdoms better qualified for the appointment than Dr. Allen Thomson, and he would also exert any private influence he might possess to further your object —

Willis promised to alter his certificate and I forward you Liston's[4] which he gave really with great cordiality and were you once in the field I think I could get him to write to some one of the Baillie Bodies with whom his opinion has weight. Clift[5] & Owen are both out of the way – so you had better apply to them by letter yourself. — I return you their documents.

Christison or Syme could have informed you 2 months ago that *I* had no intention whatever of applying for the appointment, I can see no inducement unless to have a smaller return to make for that rascally Tory Income Tax.

You say I must declare to you whether or not I would be a candidate for *Monro's Chair*[6] in case of a vacancy — Now I think all you can reasonably expect is to know *as much of my sentiments as I know myself*, and that I am most willing to tell, viz, that it would all depend on circumstances — Edinburgh might continue to drop lower, or the Patrons[7] might make arrangements which would be very onerous on Monro's Successor – or on the other hand things might get wrong in London and tho' I do not see any prospect of that, yet, as you will know, medical teaching is liable to such reverses that no one can reckon securely on the permanent success of a school. It is clear therefore that however convenient for myself it would be to know what I should do in the event of Monro resigning and however agreeable it might be to me and consistent with the feeling of sincere friendship which exists between us to inform you, yet I cannot well be expected to make up my mind now when so much depends on contingencies, though I repeat that given my mind made up you should immediately know it and I conceive you would be entitled to know it. It is precisely one of those cases on which I can decide nothing till the events which would require me to act really occur — I am satisfied in the mean time that the Patrons are committing a great blunder in taxing the new professors as they are about to do — Retiring allowances must be provided but if a strong effort were made *some other source* would be found, the Reid or Straton Bequests or the Government perhaps. It is true there will be no lack of excellent candidates for the Chairs that are not connected with practice, but the

Patrons cannot reckon on getting at all times so good a man as Alison ready to their hand when they have such a chair as the Practice of Physick to fill up — *As it is*, Home's demand of £150 beyond your father should not be listened to for a moment – he has not the slightest claim to it.[8]

Dr. R.[?] Hamilton[9] called the other morning on his way to the Continent – the same mild, intelligent, gentle creature he always was – but what an agreeable thing it was for me to see that he had undergone some sort of resurrection (one might call it) as regards his social & professional life. He is gone forth singing your praises.

I am now in the thick of Exam^n at Somerset House and hope to be through early next week. I would fain get down to Scotland in the remaining part of August and hope to make it out tho' I am wanted in London by the 1st Sep^t. I had engaged to superintend certain portions of a new Edit^n of Quain's Anat^{y}[10] but find it no easy task – the gen^l Anatomy requires to be entirely rewritten and I have hesitated so long on the threshold that I cannot with a good grace now take the play for any long time. Taylor & Walton are long suffering and most obliging, but for that very reason discretion in availing myself of their disposition is required on my part, and it is painful for me to think *how little I have accomplished* — Still I think that a few days of the northern air would put me right – and even with a view to the better attainment of the object in question would I think prove advantageous.

I should first go to Forfarshire for two or three days – but I will write you again on the matter.

I have made a fresh copy of the testimonial I gave you with a few verbal differences, some intentional, others unintended – read it over and if you have any further alteration to suggest let me know and I will be most happy to adopt it.

With sincere regards & best love to yourself and all your household.

I am your devoted
W Sharpey

Dr Allen Thomson

---

[1] Thomas Hodgkin (1798–1866), Curator of the museum at and Pathologist to Guy's Hospital.

[2] Presumably John Davy (1790–1868), physiologist and anatomist, the younger brother of Sir Humphry Davy.

[3] David Craigie (1793–1866), Physician to the Royal Infirmary of Edinburgh and editor of the *Edinburgh Medical and Surgical Journal*. He was the author of several works on morbid and general anatomy.

[4] Robert Liston (1794–1847), formerly an anatomy teacher in Edinburgh (with Syme) and Surgeon to the Royal Infirmary. He was Professor of Surgery and Clinical Surgery at University College London from 1835.

[5] William Clift (1775–1849), former apprentice to John Hunter and Conservator of the Hunterian Museum at the Royal College of Surgeons, London.

[6] I.e., the Chair of Anatomy at Edinburgh.

[7] The Town Council of Edinburgh, which had the right to make appointments to University chairs.

[8] Home was apparently demanding a larger pension from the University than John Thomson had requested upon his retirement from the Pathology Chair in 1841. In the event, Home accepted £100 more than John Thomson received from William Henderson, the new Professor of Pathology. Allen Thomson had a personal interest in these proceedings because he was obliged to pay half of Home's pension, while Alison supplied the remainder. See: Edinburgh University MSS, College Minutes, 1838–44, vol. 6, pp. 439–46, Senate Meeting of 30 September 1842.

[9] Possibly Robert Hamilton (1794–1868), who graduated MD in Edinburgh in 1815 and was President of the Royal Medical Society in 1814.

[10] I.e., J. Quain, *Elements of Anatomy*, 5th ed., 2 vols., London, Taylor and Walton, 1843–8.

21

London 13 Oct 1842

My dear Thomson

Excuse me for writing on this scrap of paper for there is none other within my reach —

It is almost too late to congratulate you on your success but not too late to say that it has given the greatest satisfaction to every one whom I have heard express themselves on the subject.

I should have written sooner to say that so far I can order your tubes you must give me the extreme breadth of the part of your eye piece *which is to be introduced into the tube* – if they are as my french ones you will find that the ring which holds the field glass or large glass of the eye piece is a little wider than the part immediately above, you must give me the diameter of this. My brother suggests that a hole cut in a card to the size and sent in a letter would be the best measure –

I fear I cannot keep my promise of presenting you with an adjustment, for my brother has expended much more work on them than I intended and has only made three in all – they are made of a stout circular mahogany base supported on three very low knobs – two uprights about 9 inches high strongly morticed into the base – a cross bar of brass piping from the top of one upright to the other, a piece of tube of about 3 inches long fixed vertically thro' the middle of the transverse bar – a long tube containing the glasses passed thro' the short tube & fitting with a sliding motion –

Stage consisting of a plate of sheet brass with a round hole in it, fixed across the uprights about 3 1/2 inches from the bottom, it is shored (permanently) into a saw rut in each upright which makes a very tight fitting – The mirrors are placed below and finished in the usual way – Now I intended that the mirror should be a much more simple affair – and that the cross bar which holds the fixed tube should be merely of wood. My Brother says that the tube may be very firmly fixed into it by means of a mixture of sealing wax & pounded brick melted on the Fire – I find the fitting will answer very well for *exhibition* as intended, for you can by giving the tube a screwing motion within the other find the focus with tolerable ease and once found it is not liable to be deranged by the inspector – It will not be so well adapted to recommend to students as a *working* microscope. For low eye pieces and opaque objectivity – like most of your best preparations – it will answer admirably.

The wooden uprights are a little [drawing here, see figure 2] elastic and yield somewhat when one adjusts the instrument but it is by no means a *shakey thing* – once the focus is adjusted it is as firm as a rock — I think a *broad brass stage* with a hole in it is better than any thing else, for when you lay a good big piece of plate glass on it for your object you can move it about almost as easily & firmly with your fingers as with the screw stage — the stage may be made to project more behind so as not to cut off light from the mirror – & being broad [?] it will hold a frog when you show the circulation –

I have ordered glasses for you but have not yet received them — I will however send you down what you desire from my own store in the mean time. Mr. Halley[1] who is to be in Edin. this Winter will take them with him, he goes by the Steamer on Saturday.

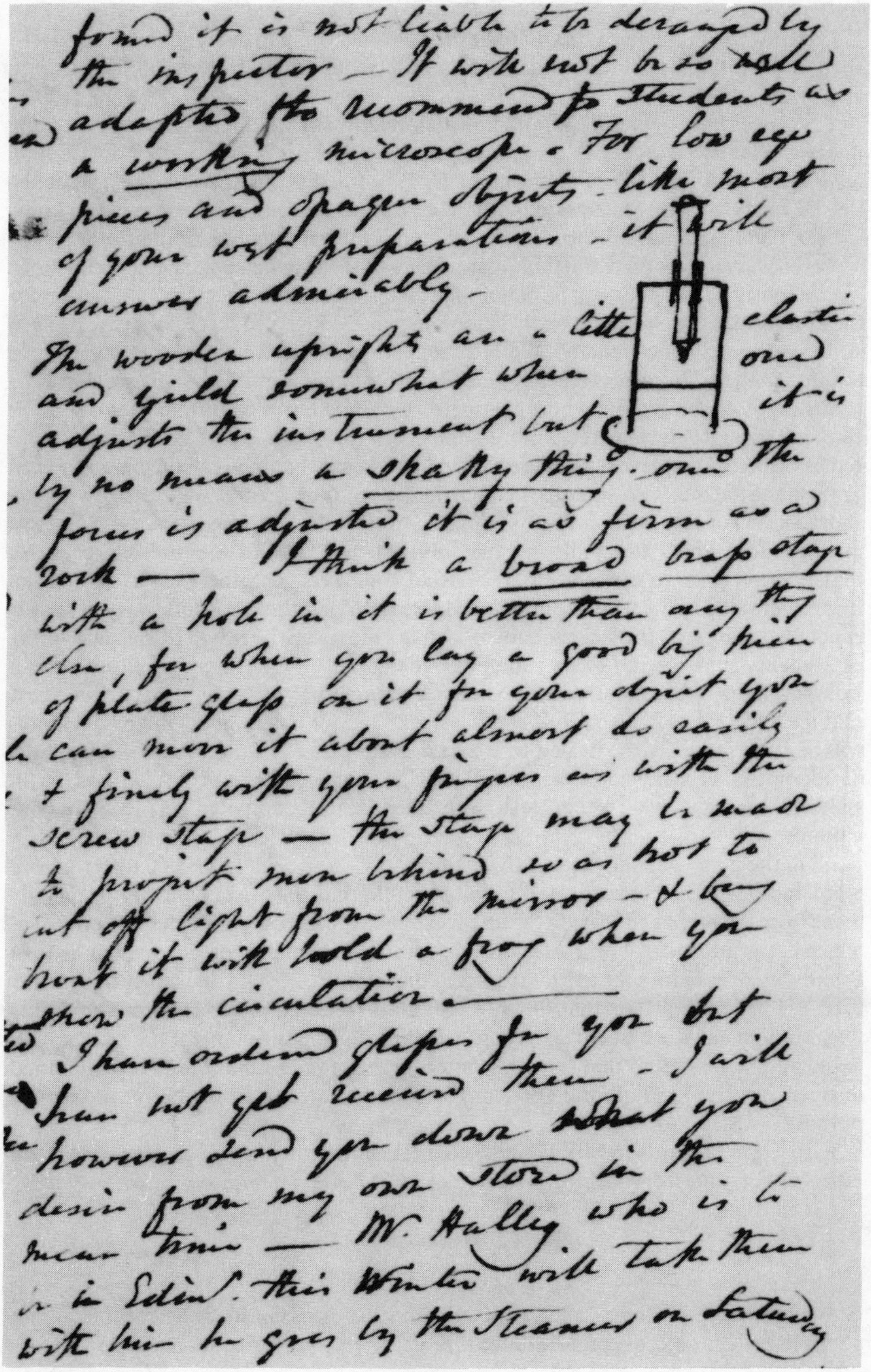

found it is not liable to be deranged by
the inspector — It will not be so well
adapted to recommend to students as
a working microscope. For low eye
pieces and opaque objects. like most
of your wet preparations - it will
answer admirably.

The wooden uprights are a little elastic
and yield somewhat when one
adjusts the instrument but it is
by no means a shaky thing. once the
focus is adjusted it is as firm as a
rock — I think a broad brass stage
with a hole in it is better than any thing
else, for when you lay a good big piece
of plate glass on it for your object you
can move it about almost as easily
& finely with your finger as with the
screw stage — the stage may be made
to project more behind so as not to
cut off light from the mirror - & by
that it will hold a frog when you
show the circulation.———

I have ordered glasses for you but
have not yet received them — I will
however send you down what you
desire from my own store in the
mean time — Mr. Halley who is to
is in Edin.[?] this winter will take them
with him he goes by the Steamer on Saturday

Figure 2.
33

I will not forget the mica tho' I dont exactly know where to get it perfectly good — I shall find another opportunity for the tubes I have no doubt, tho we can ill spare you any of our students.

The classes have opened here, and I must decidedly say we shall have a short-coming in new students, according to present appearances I take it that in our School we shall have about a *fifth* deficiency on the numbers of last year – this applies to new students, the old will be certainly not less in N^o than before, but unluckily the new men you know are the Pabulum Vitae. From what I can learn of other London Schools at least the *larger ones* such as Guys & Bartholomews I believe we have reason to be thankful – our *relative* position will I suspect be better than last year – but that is poor consolation for the absolute loss — As to the small Schools in general I hear of several that are *bad bad* Aldersgate nearly knocked up – Webb Street *had been* just given up in the course of the Summer and St Thomas's with all their flourish of drums and trumpets and with the *debris* of the Webb Street[2] whom they receive on like terms as former pupils of their own, are likely to have scarcely more pupils than they have got of Lecturers — As to Kings College, Damill told me there was an increase in what they call their *matriculated* or regular Students who enter to their curriculum of four years & wear the gown but a short coming in *occasional* Students or such as dont enter to all the classes — what this means I dont know but I suspect they are not well off tho' I can easily conceive that *mere connexion* may get the usual numbers together in a school so inconsiderable — The whole is I think an indication that there is a deficiency of supply generally seeing that the large schools which are most dependent on the general influx of Students are likely all to be sufferers — I sincerely hope it may be different with Edinburgh tho' I fear that the same cause will tell against you — In the mean time now that the Session has gone on for near two weeks and we have a good grasp how things will be I am perfectly reconciled to it and it is curious as I now write to feel as little disappointed as last year – the consolation is that matters are not so bad as in 1839–40 – & that was a contrast with the capital campaign in 1838 – pejora vidimus. —

I find the plate glass bottoms answer admirably – and really I believe the Gold size and lamp black the best cement after all — I find that the air bubbles very generally come from the preparation itself especially if put into spirit of diff^t strength from that in which it has previously been — Air also is apt to get in unless you lay on your cement round the cover before the spirit dries up to the joining of the lid and cut surface of the tube — I believe however that Drake's thin glass (selected pieces I mean & tolerably flat) would make the best covers, it is so yielding that the shrinking of the spirit would do no mischief — I find that the glass cutters here use *sand* & not the [. . .] of diamond dust – it is a firm kind of sand got near Croydon in Surrey that is chiefly used for the purpose —

With sincere regards to Mrs Thomson & best wishes for yourself both generally and especially in your approaching session — I am My dear Thomson
Yours affection^{ly}
W Sharpey

I have given a Mr Williams[3] an introduction to you – he is not likely however to attend physiology – his object being to graduate & attend the more practical subjects — He is a little peculiar & odd but a very intelligent and *most excellent man* his father is Sir Somebody Williams the late Mayor of Shrewsbury.

WS

P[lease] T[urn] O[ver]

You ask about the change in the Inspectorship of Anatomy – I really know nothing further than this – that a Committee of the College of Surgeons consisting of Stanley[,][4] James Arnott[5] & South,[6] had been sitting during the Summer on the subject of the Supply of bodies to the Anat. Schools & that they recommended (or rather Stanley who was their *primum movens*) recommended a change of measures – and this I presume Sir J. Graham conceived to involve a change of men — This *is all I know*, as to Somerville's partiality &c. &c. – it is all nonsense and worse than nonsense, but Sir J. Graham failing to unseat him through means of this late Committee of inquiry (I mean the ordeal which Somerville had to pass through *last winter*), and being determined to get rid of him & please Brodie just thought he might do that by hook which he could not do by crook, and so gave him his leave — I understand Christian & Wood were written to[,] to recommend a person for Edinburgh (by Brodie? or thro' his instrumentality no doubt) and Mr. Wood very naturally recommended Andrew – who (if he would cut one of his recurrent nerves) will do well enough.

W.S.

[1] Alexander Halley (1824–75) graduated MD in Edinburgh in 1844. He practised in Leeds and later in Harley Street, London.

[2] The Webb Street private anatomy school was incorporated into St Thomas's Hospital in 1842 upon the appointment of its proprietor Richard Dugard Grainger as lecturer on general anatomy and physiology at the hospital.

[3] I.e., Philip Henry Williams (1821–72), who graduated MD in Edinburgh in 1843 and later practised in Worcester.

[4] Edward Stanley (1793–1862), surgeon and lecturer in anatomy at St Bartholomew's Hospital. He was elected a Life Member of the Council of the Royal College of Surgeons in 1842.

[5] James Moncrieff Arnott (1794–1885), surgeon to the Middlesex Hospital; see also note 44.1.

[6] John Flint South (1797–1882), surgeon and formerly lecturer on anatomy at St Thomas's Hospital. He had been a member of the Council of the Royal College of Surgeons since 1841.

22

London 4th May 1845
35 Gloucester Cresent
My dear Thomson,

Your friend Mr Ackland[1] faithfully delivered his charge on Friday after noon, but as he was about to start for Edinburgh again the next day he remained with me little more than a quarter of an hour — When he affords me the opportunity (which he has promised to do) I will be most happy to afford him all the information and show him any attention in my power; but I have told him that he is likely no where to meet with useful advice as to teaching more readily than from yourself.

35

Pray accept for yourself and for Goodsir[2] my best thanks for your kind present – I have looked over the things with Ellis and Potter[3] and am much pleased with them. The *tuberculous lung* is both a useful and handsome specimen – not to speak of the others in particular.

Perhaps too you will have the kindness to repair an omission of which I have been guilty towards Goodsir, and thank him in my name for the copy of his Researches[4] which he has been so good as send to me – I have read through the greater part of the book & with much pleasure —

I expected no other reception for Quain's Pamphlet[5] — It may be open to the charge of minor inaccuracies, but the real cause of offense in certain quarters is its truth. Of course it is unpalatable to the *old* Faculty who resisted pertinaciously for 15 years the opinion of all judicious men without its pale in the matter of preliminary education, and who admitted Midwifery & gen*[l]* Pathology into the medical curriculum only on compulsion. It may be abused too by some who now find it convenient to take part with them, and who in my opinion confide more in their good intentions than past experience warrants. But really after an interchange of letters with Syme in which there was some anger (on both sides I freely admit) I am not disposed to run any further risk — Not a word is said in the pamphlet to disparage the instructions given in Edinburgh and for the best of reasons, because the writer of it holds opinions incompatible with any such disparagement; but that does not prevent Syme from telling me that we fail to make good surgeons, and that because we are not aided by the influence of *our character*, — he casts up to me a foolish step of one of my colleagues in presiding at a meeting of students, as if the Edinb. Faculty invariably kept its members from indiscreet acts. This in answer to a letter in which I reiterated an opinion I have never concealed viz that the Faculty were to blame for neglecting the preliminary education of their graduates, and in which I ventured to criticize (too sharply perhaps) the proceedings of Dr. Christison[6] as representative of the University to the Government and writer in the Edinb. Review. — This for yourself and intimates (not for gossip). I have since written Syme again and in a cooler mood and I admit that my first letter was all the stronger in consequence of the tone in which Syme had thought proper to write to Mr. Quain and in consequence of a letter from Christison to Tweedie[7] which was read to me immediately before.

There is no hurry with the Embryo — send it when quite convenient and by a sure hand — Koelliker[8] from Zurich has been here some time and is now about to return home, have you seen his memoirs on the development of Cephalopods?[9]

I am most obliged to Goodsir for his book but I am not sure if it was well advised to republish all the papers it contains — He is an excellent observer and sound headed man, but too anxious to gain a reputation as a generalizer in science too prone to aim at establishing general propositions – the great aim no doubt of all science but not to be done rashly.

The consequence is that much will require to be unsaid. He is too fond also of employing new phrases – and not careful to avoid mixing inference and even assumptions with his expositions of facts. The paper of centres of nutrition[10] was well as a passing contribution, but scarcely deserves its present place — That on the structure & nutrition of bone[11] is altogether wrong. — I have great doubts too as to the essay on Cystic Entozoa,[12] and that on reproduction of lost limbs of Crustacea.[13] Of course this is "privatissime" for the book is *there* and the above remarks can now do no good. Moreover I know no one likes criticism and I have a great esteem for Goodsir. — With best regards to Mrs Allen & the household at Moreland Law

Your sincere friend

W Sharpey

[1] Possibly Henry Wentworth Acland (1815–1900), who was a student in Edinburgh from 1843 to 1845. Acland became Professor of Clinical Medicine at Oxford in 1851 and served on the General Medical Council.

[2] John Goodsir (1814–67), Curator of the Museum and Demonstrator in Anatomy at the Royal College of Surgeons, Edinburgh. He was appointed Professor of Anatomy upon Monro's retirement in 1846.

[3] George Viner Ellis and John Phillips Potter, Assistant Demonstrators in Anatomy at University College London.

[4] I.e., John Goodsir and Harry Duncan Spens Goodsir, *Anatomical and pathological observations*, Edinburgh, Macphail, 1845.

[5] Richard Quain, *Observations on the education and examinations for degrees in medicine, as affected by the new Medical Bill, with remarks on the proposed Licensing Boards, the Society of Apothecaries, the Registration of Medical Practitioners and the constitution of the "Council of Health"*, London, John Murray, 1845.

[6] See *Edinb. med. surg. J.*, 1845, **63**: 159.

[7] Alexander Tweedie was at this time examiner at University College London.

[8] Rudolf Albrecht von Kölliker (1817–1905), German histologist. He had held the post of Associate Professor of Physiology and Comparative Anatomy at the University of Heidelberg since 1844. He had *previously* been at Zürich.

[9] *Entwickelungsgeschichte der Cephalopoden*, Zürich, Meyer and Zeller, 1844.

[10] 'Centres of nutrition', in Goodsir and Goodsir, op. cit., note 22.4 above, pp. 1–3.

[11] 'The structure and economy of bone', ibid., pp. 64–7.

[12] 'Of the anatomy and development of the cystic entozoa', ibid., pp. 79–103.

[13] 'The mode of reproduction of the lost parts in the crustacea', ibid., pp. 74–8.

23

35 Gloucester Crescent, London

11th June 1845

My dear Thomson,

I do not know whether you may have any young men of merit to recommend for the office of Conservator to Edinb. College of Surgeons; but let this be as it may I have promised to write you to obtain a favorable consideration for the application of a young man who was formerly a pupil with Mr. Hamlin Lee. He is an intelligent, gentlemanly person, remarkably steady and industrious; and he has devoted himself very much for some time to comparative anatomy and palaeontology as assistant to Dr Mantell.[1] He is well acquainted with the use of the microscope for he has devoted much attention to the study of the infusory animalcules and other minute organisms – I think altogether you would be pleased with him.

I met Dr Patrick Newbigging[2] ten days ago at your friend Mr Murray's he tells me you have moved to Hope Street, which is not only a pleasant situation but *convenient* for Morland. I hope all the inhabitants of both places are well — I heard from Dr William [Thomson] the other day, about an edition of Wm Hunter's works contemplated by the Sydenham Society. Talking of Wm Hunter reminds me (by force of contrast) of Dr Robert[3] here and his uterine nerves[4] — Mr Beck[5] has at length sent his paper to the Royal Society, Brodie[6] presented it. Lee will be quite floored at least in every body's estimation but his own. John Reid was here a short time ago and saw Beck's preparations —

The St. Andrews professors desire to give up the practice of granting Medical Degrees, provided they can get a compensation. It's no use attempting to dragoon them out of it when a hundred pounds or two a year will buy them off — Reid speaks very sensibly on this matter, but complains of the very unwarrantable statements put forth in the Edinburgh Review,[7] with which he found Sir James Graham had been most industriously stuffed. I cannot help still thinking that the accredited representative of the University of Edinburgh should have informed himself better, if he could not refrain from making attacks on other places.

Dr Wm Macdonald dined with me one day last week, I am sorry for him poor fellow. He means to work in Canada at last, unless Sir Robt Peel selects him as specially qualified for a Professor in one of the new Irish Colleges.

With kind remembrances to Mrs Allen

believe me

My dear Thomson

Your sincere friend

W Sharpey

[1] Gideon Algernon Mantell (1790–1852), physician and geologist, was resident in London after 1839.

[2] Patrick Newbigging graduated MD in Edinburgh in 1834 and became FRCSE in the same year. He was physician to the John Watson's Institution.

[3] I.e., Robert Lee (1793–1877) obstetrician and author of works on midwifery and gynaeocology.

[4] *On the nervous ganglia of the uterus*, London, R. and J. E. Taylor, 1841; reprinted from *Phil. Trans. R. Soc.*, 1841.

[5] Thomas Snow Beck, 'On the nerves of the uterus', ibid., 1846, pp. 213–35. Sharpey was accused by the *Lancet* of colluding with Beck to demolish the views expressed in Lee's 1841 communication. See Taylor, op. cit., note 2 above (*Introduction*), pp. 244–5.

[6] Benjamin Collins Brodie (1783–1862), Sergeant-Surgeon to the Queen and Surgeon to St George's Hospital. He had been a Fellow of the Royal Society since 1810.

[7] See [R. Christison], 'Medical reform', *Edinb. Rev.*, 1845, **81**: 235–72.

24

London 29th Oct. 1845

My dear Thomson,

Many thanks for your letter and for your taking the blame of the long interruption to our correspondence on yourself, but honesty compels me to say that the fault lay with me. Your commission shall be executed the moment my brother Alexander, who is at present in Manchester, arrives in London which will be in a day or two.

Our school has opened and in respect of new entries is the same as last year not one above and I believe not more below. This is not saying much, but from what I can learn of the schools generally I believe we may be thankful.

You congratulate me on the prospects of my general anatomy[1] – I wish I could deserve your congratulation. The publishers are determined to bring out the first half of the book in two or three days, but my contribution is barely begun — We have adopted a different paging for the "general anatomy" and there will be little more than the preliminaries now published – viz the generalities in the vital and physical properties and development of the textures and in the blood – so you must not judge it till it is all out, in fact the publication is hastened rather on account of the descriptive anatomy of the bones joints and muscles to serve our students along with Ellis's book[2] as a text book for the Session.

I could have offered a classification of the textures, but at present this is all guess work. Classification is not needed in order to *manage* the subject as in the *mutitude* of objects of Botany or Zoology, it can only be useful as showing the mutual relation of the textures and really this is too uncertain — In my lectures I am in the habit of giving a classification at the *end* when the relations can be made more intelligible.

The only excuse I can give for the preliminary account of the *more frequently occurring* chemical compounds of the animal body – is simply *convenience*. It is difficult to speak of "osmazone" and the varieties of "fat" etc. *comfortably* without having explained what you mean; the only other alternative was to leave *them out altogether*, and presuppose the knowledge; this I grant is the logical-philosophical view to take but practically the case assumed a different aspect; and by freely employing the smaller form of type on such occasions (for we have adopted a larger and smaller) I think I have kept things in their due subordination which is a greater point in an elementary treatise. To stop your mouth I shall cause a copy to be sent to you.

You once asked me about Martin Barry's spiral.[3] Last summer [. . .] he showed me muscular fibres from the turtles [and] frogs heart which he called double spirals, they had this appearance [see figure 3] which was obviously due to two rows of disks placed a little slantingly – he then brought things like this [drawing here] but these seemed to be threads of cellular tissue — lastly he produced one or two specimens of his own & one which he fished out among mine (of the growing fibre in the tad-pole), mine was not very conclusive though he persisted in thinking so – his own I must freely confess imposed on me; they had this look [drawing here] quite like an opened single spiral, quite the aspect of a *continuous thread*, and obviously part of the muscle — the only difficulty with me was, that only two or three of them could be found after all his searching, and this ought to have put me on my guard, and moreover they appeared far more distinct in his microscope than with mine, – *but there they were* & I could not see any fallacy; it occurred to me therefore that the filaments might originally be a spiral thread which got afterwards broken up into disks or rings as Mohl[4] supposes to happen in the formation of annular ducts in vegetables – & that this might account for their being only exceptional in their occurrence for without doubt the muscular filaments generally were composed of series of disks or segments. *So I was deceived* and I will never again give an opinion on specimens that I have not leisurely examined, at home and by myself, repeatedly; that is, in doubtful cases. At length on looking at a specimen of pig's muscle in one of Powells microscopes the following appeared – two fibrillae or filaments lay close together [drawing here, see figure 4] *when in focus* they appeared made up of light and dark segments which alternated in position in the two

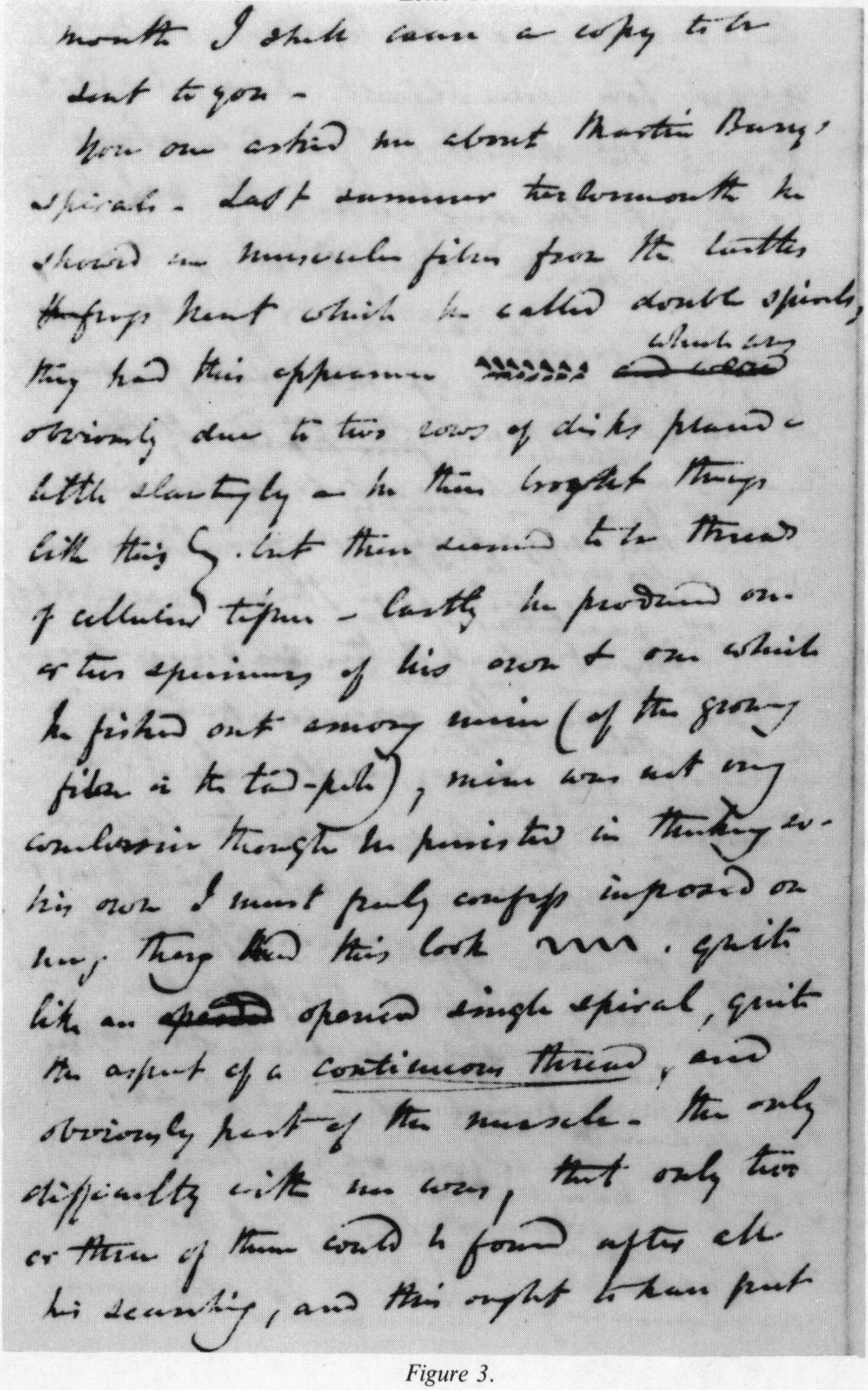

month I shall cause a copy to be
sent to you —

You one asked me about Martin Barry's
spirals. Last summer ten bermouth he
showed me muscular fibres from the turtles
the frogs heart which he called double spirals,
they had this appearance ~~~~~ ~~~~~ which are
obviously due to two rows of disks placed a
little slantingly as he then brought things
like this ⟨⟩. but there seemed to be thread
of cellular tissue — lastly he produced one
or two specimens of his own & one which
he fished out among mine (of the growing
fibre in the tad-pole), mine was not very
conclusive though he persisted in thinking so —
his own I must freely confess imposed on
me; they had this look ~~~ . quite
like an opened single spiral, quite
the aspect of a <u>continuous thread</u>, and
obviously part of the muscle — the only
difficulty with me was, that only two
or three of them could be found after all
his searching, and this ought to have put

Figure 3.
40

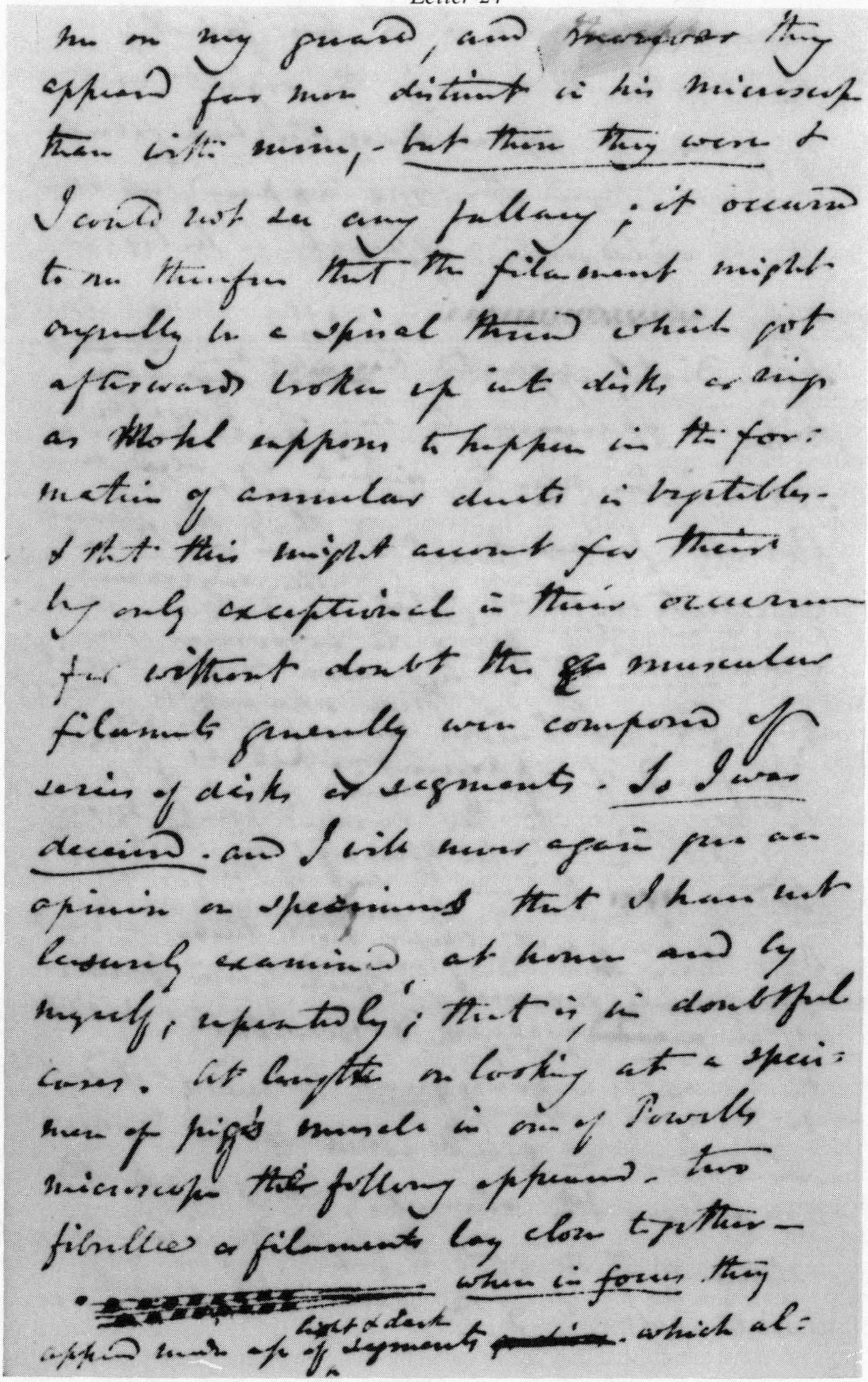

me on my guard, and moreover they
appeared far more distinct in his microscope
than with mine, - but there they were &
I could not see any fallacy; it occurred
to me therefore that the filament might
originally be a spiral thread which got
afterwards broken up into disks, or rings
as Mohl supposes to happen in the formation of annular ducts in vegetables.
& that this might account for their
by only exceptional in their occurrence
for without doubt the muscular
filaments generally were composed of
series of disks, or segments. So I was
deceived. and I will never again give an
opinion on specimens that I have not
leisurely examined, at home and by
myself, repeatedly; that is, in doubtful
cases. At length on looking at a specimen of pig's muscle in one of Powell's
microscopes the following appeared - two
fibrillae or filaments lay close together -
                                    when in focus they
appeared made up of light & dark segments which al-

Figure 4.
41

fibrils, on slightly moving the object out of focus the segments of the two fibrils coalesced and the two appeared as one thread coiled into a spiral or helix [drawing here, see figure 5.]

I immediately requested the Quaker to submit his specimens to one of Powells best instruments, which we did in Powell's presence, and tho' the result was not so clear as in the instance stated, it sufficed to convince me that these two or three examples Barry had found of apparent spirals were really owing to the cause just mentioned. [drawing here] [T]his was the sort of thing, and then it was quite clear that there were *two fibrils* present, which it is but fair to say, Barry had previously held to be the case. The appearance is most deceptive when one fibril comes to lie somewhat over the other & when you see the segments of the one through the interstices of the segments of the other; the slanting direction greatly helping the illusion.

But since then my eyes have been a little opened to the quality of the Doctors investigations — I had occasion for a particular object carefully to peruse his extraordinary papers on cells in the Phil. trans. and to *re*peruse all his older ones on the ovum.[5] They are full of extravagances. I never gave in to his notion of the generation of a filament in the blood corpuscle. The appearance he described always appeared to me to be due to corrugation — And I must tell you a little anecdote, which at the time I rather chose to speak of only to friends as the poor Quaker was being worried by that cankered little cat Wharton Jones and I thought there was quite enough to overwhelm him without my lending a hand. The story was this. Barry one day brought a doe rabbit to the College to show me the wonderful spontaneous movement of the blood corpuscles which he had discovered *at a certain stage after* the *coitus*, and which *of course* Dr. Carpenter[6] had adduced as proof of an inherent vital contractility of the red disks &c. &c. &c. The beast was killed and opened and the Dr. carefully *scraped* a little blood from the end of the fallopian tube and surface of the ovary, then displayed it under his microscope (for he carried his famous Schilck in a blue camlet bag through the streets) and desired me to look and wonder. I looked and sure enough there were the globules dancing and behaving themselves in the most ridiculous way like the fellows in a pantomime, indeed just as they are elaborately represented in their different phases in the Philosophical Transactions. But I had not looked a minute when I saw clearly the whole was owing to cilia – and indeed I could see the films of ciliated epithelium with the cilia moving on them, and it was only where they were seen that the blood disks were dancing. The epithelium was transparent & Barry could not see it or the cilia — I called Marshall[7] (the young man of [the] H[unterian]. Museum) & he saw them plainly enough, the Doctor was not charmed. I pointed out that where the field was quite free from cilia and quite transparent the corpuscles were quiet — No says the Dr. I once myself thought it was an affair of cilia but I satisfied myself it was otherwise — Depend on it says I you'll find the motion in any she rabbit you choose whether she has had the benefit of the male or not, but there is a frog in the next room, let us put some frog's blood on the same spot as the rabbit's blood and see what will happen. The frog's blood corpuscles are mixed with the other on the field of the microscope, and they are also kicked about like the others, though not so much distorted and disfigured, for they are larger and stiffer — The Doctor does not see it at first, but at last admits that they are served in the same way – and in the end he turns to

terminated in position in the two fibrils, on slightly ~~——~~ moving the object out of focus the segments of the two fibrils coalesced and the two appeared as one thread coiled into a spiral or helix —

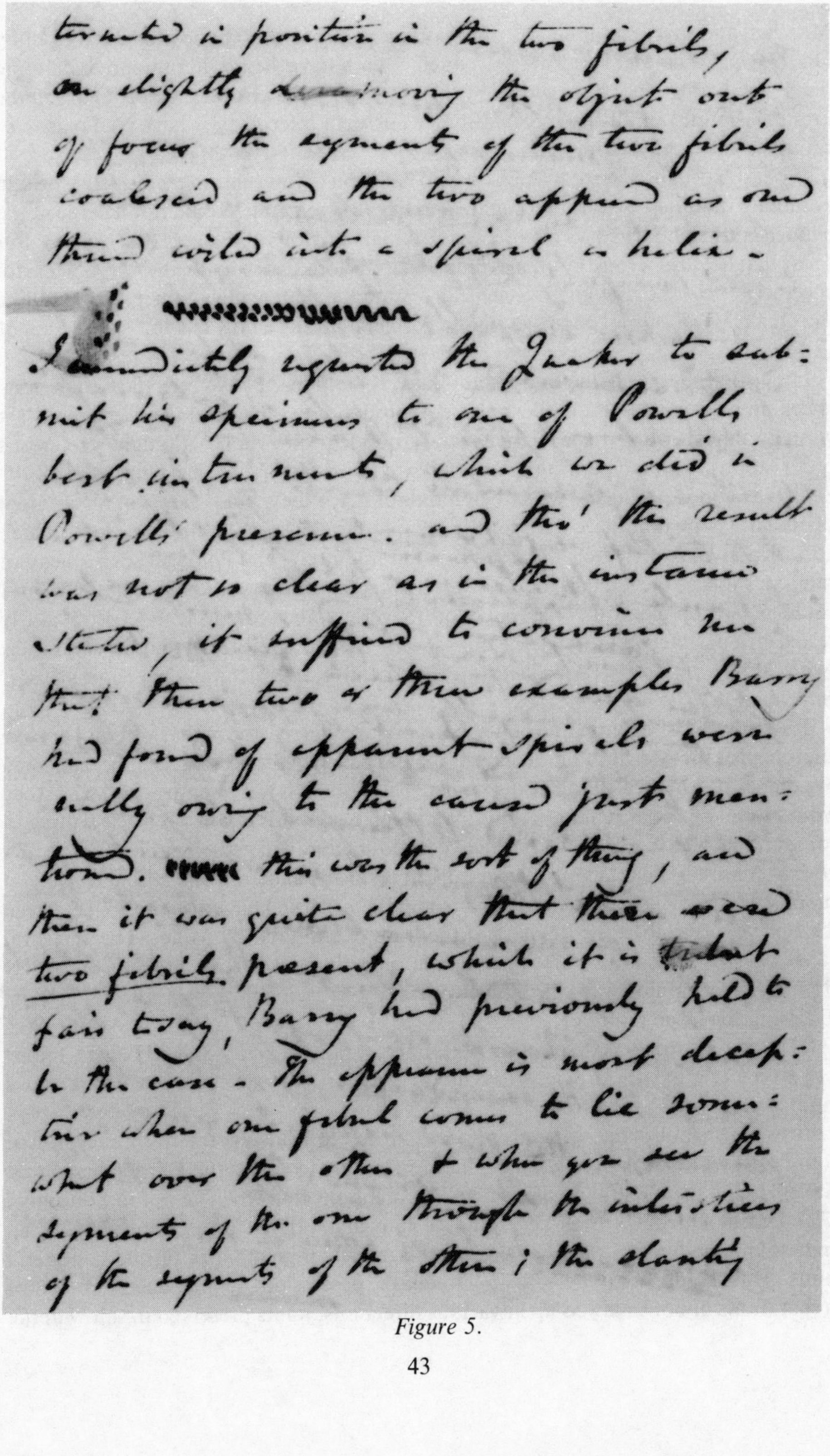

I immediately requested the Quaker to submit his specimens to one of Powell's best instruments, which we did in Powell's presence. And tho' the result was not so clear as in the instance stated, it sufficed to convince him that the two or three examples Barry had found of apparent spirals were really owing to the cause just mentioned. ~~——~~ this was the sort of thing, and then it was quite clear that there were two fibrils present, which it is but fair to say, Barry had previously held to be the case — The appearance is most deceptive when one fibril comes to lie somewhat over the other & when you see the segments of the one through the interstices of the segments of the other; the slanting

Figure 5.

Marshall and says, that if he was to be corrected he was very happy to be put right by Dr. Sharpey – & there the matter ended; but I have heard it rumoured that the phenomenon has since been shown by the Dr. or his adherents in the Borough,[8] on the original footing, and if this information be correct I infer he may have seen reason to return to his published opinion again.

What startled me was, that a person who has been so long and so exclusively working at minute inquiries, who is a draughtsman into the bargain; should not have seen the ciliated epithelium which to my eye was strikingly obvious at the very first sight — But I fear that he is led wrong by an inordinately ambitious desire to *discover*; he is a fearful spectacle of morbid craving for scientific distinction, and while in other respects I like the Quaker, you cannot conceive what a perfect loathing I now feel for that species of vanity.

I hope that the good Doctor does not know it, but I fear I have been the first to put a check on his papers in the Philosophical Transactions — He read a long memoir last winter, almost wholly speculative – and founded on what I thought erroneous observations, with a repetition of or reference to his old errors. The committee of the Society I suppose wished to wash their hands of it, for they sent it to me for an opinion; I could not deny that there were some probable enough speculations in it, some which accorded at least with facts observed by other inquirers; and amongst such a heap of random shots as the Quaker fired it was possible that some would tell in the end; but I could not recommend it for publication. Several statements in it indeed obviously rested on a misinterpretation of the things seen. The Doctor learned that the committee (who adopted my report) declined to recommend it and withdrew his paper. *This is entre nous.*

The little bit of examination duty Mr Naylor referred to, is that of comparative anatomy for the Assistant Surgeoncy in the E. India Co' service that is to be conferred by competition in the College next summer. The College is too poor to afford what they would deem a proper remuneration to a man like you, but if you would be willing to come and spend a few days with us, I have no doubt they will get something over paying your expenses, besides having the pleasure of laying us under an obligation to you — I suppose your article ovum wont be printed till then, & if so I have an embryo – between the one you described (of Reids) and the little one figured by Miller. The vertebral canal has just closed but burst open again, or perhaps it has never been completely closed — The amnios was entire but I opened it as well as the yolk sac.
[Drawing here, see figure 6]
I still intend giving some description of it, but if it cant be done sooner I believe I must let you figure it in your articles — So if you come up I promise you it will make your mouth water —

But my own mouth has watered ever since I received your very interesting account of your brother William's excavations – the moment I read your letter I remembered that Haller[9] in the sketch he gives of his own life, mentions that when he was in London in 1727 Dr Ja^s Douglas[10] was very kind to him and offered to associate him in an undertaking he himself was then engaged in namely a great work on the bones and joints. Haller adds that he saw in Douglas's possession an immense number of preparations of bones sawed up in various directions, joints preserved in salt water &

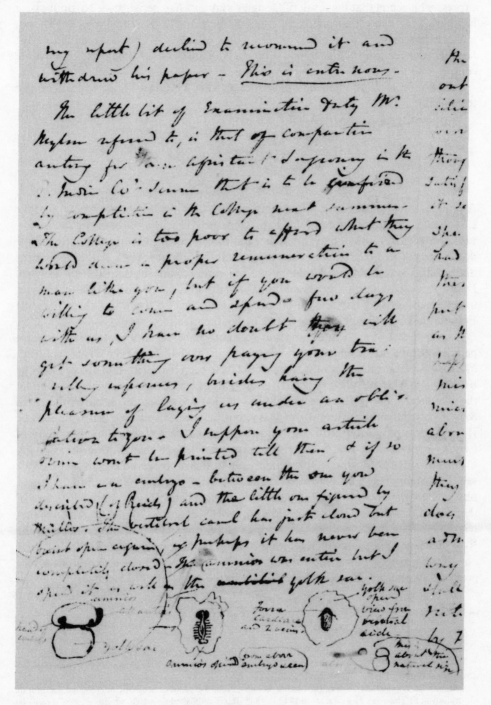

Figure 6.

expresses his regret that the result of Douglas's labours had perished — No doubt your brother's very interesting discovery explains their fate — possibly [William] Hunter has purchased them with others of Douglas's effects[11] — One is reminded of the famous resurrection of Eustachian's tubes, by Lancisi, after being buried thro' [a] century and a half —

I did not reckon on so long a fit of scribbling when I began – I hope you will be able to read it

With best regards to Mrs Allen & all the rest I am

> My dear Thomson
> Your sincere friend
> W Sharpey

As you refer to the Alchemist[12] I may just remark (tho' the thing is stale now) that the London Chemists (I dont mean Graham) think he is cracked. *He* came up last winter with a letter of introduction to Damill begging him to undertake an investigation of the doctrines; but he hovered about the class room for about a week & then presented himself and it was observed that he seemed rather thankful than otherwise when Damill excused himself from meddling with the affair — I wish Graham had not meddled either, and had let the all sufficient wisdom of the thirty three [i.e., the Edinburgh Town Council] bring about its legitimate effects.

I never heard of such infatuation as seems to have possessed otherwise sensible people – Would not *common sense* have told them that had there been an atom of truth in Brown's extraordinary results, they would have been blazed abroad all over Europe by now? Did [Humphry] Davy need to go begging of chemists to verify his discovery of the alkaline metals? — It is true it might be said that Brown was an obscure young man & in a very different position from Davy, but he was backed by Christison and the Royal Society of Edinburgh, though Christison seems to me now to show symptoms of *backing out* — I dont find fault with a mistake. The best men will be occasionally mistaken, it is the obstinate, conceited, self sufficient perseverence in it – and the senseless outcry that Chemists wont attend to his discoveries and above all the stupid obstinacy of the Town Council;[13] the only legitimate conclusion from the disregard of the alleged discoveries, is that they are not worth regarding – but I dont doubt that trials have been made in most laboratories of Europe in [. . .] of the sanction of the Royal Society – but trials which soon showed it was unnecessary to go farther. — But I am on dangerous ground, and I daresay Brown may not only be a clever but a very good fellow tho' he *deceived* [?] *my brother* in his dread of anticipation.

WS

[1] I.e., Sharpey's contribution to the fifth edition of Quain's *Anatomy*, op. cit., note 20.10 above.

[2] George Viner Ellis, *Demonstrations of anatomy; being a guide to the knowledge of the human body by dissection*, London, Taylor and Walton, 1840.

[3] Martin Barry (1802–55), one of the British pioneers in the use of the achromatic microscope, lecturer in physiology at St Thomas's Hospital from 1843 to 1845. Barry published an article 'On fibre', *Phil. Trans. R. Soc.*, 1842, Pt. 1, pp. 89–135, which held that fibres were made up of spiral filaments.

[4] Hugo von Mohl (1805–72), German microscopist who did important research on the structure of the vegetable cell.

[5] 'Researches in embryology. First series', *Phil. Trans. R. Soc.*, 1838, Pt. 1, pp. 301–41; 'Researches in embryology. Second series', ibid., 1839, Pt. 1, pp. 307–80; 'Researches in embryology. Third series: a contribution to the physiology of cells', ibid., 1840, Pt. 1, pp. 529–93; 'On the corpuscles of the blood', ibid., pp. 595–612; 'On the corpuscles of the blood—Part II', ibid., 1841, Pt. 1, pp. 201–68; 'On the corpuscles of the blood—Part III', ibid., Pt. 2, pp 217–68.

[6] William Benjamin Carpenter (1813–85), Professor of Forensic Medicine at University College London and author of numerous textbooks. A letter from Carpenter to Barry is printed in ibid., pp. 228–9.

[7] John Marshall (1818–91), Demonstrator of Anatomy at University College London.

[8] I.e., the Borough Hospitals, Guy's and St Thomas's.

[9] Albrecht von Haller (1708–77), Swiss anatomist and physiologist.

[10] James Douglas (1675–1742), London anatomist and man-midwife.

[11] The referent of these statements is obscure. They seem to imply that William Thomson had identified some of the preparations in the Hunterian Museum in Glasgow as derived from Douglas's collection. Thomson had in 1839 conducted a survey of the Hunterian Museum: see Coutts, op. cit., note 28 above (*Introduction*), pp. 515–6.

[12] I.e., Samuel Brown (1817–56), a chemist who held unorthodox views on the isomerism of certain substances.

[13] The Edinburgh Town Council had, in 1843, refused to appoint Brown to the Chair of Chemistry.

25

London, 27th Dec[r] 1845
35 Gloucester Crescent
My dear Thomson,

I only got your letter yesterday – as I did not happen to call at the College on Christmas day.

The private nature of your communication precludes me from consulting with any one here, and I scarcely regret it, because among my few intimate acquaintances here it would be difficult to find a fit counsellor in the circumstances — Imagine yourself for a moment away from Edinburgh and then you will be the best friend I can advise with.

I feel comfortable and satisfied here in the mean time, and I must say that residence in London as such – though in a humble quiet way such as I am in, has much to reconcile one to the absence of old friends and early associations, and assuredly unless Edinburgh were to me very different indeed to what it would be to a stranger I would scarcely change my present position for mere pecuniary advantages were they much greater.

My income from my class last year was £530, it will not be less this year – from this sum must be deducted £50 which I contribute towards the pay of the Demonstrators — Then I receive £175 as Examiner at the University of Lond. Thus £530 − £50 + £175 = £655. Such is my present emolument here — From your account I gather that in Edinburgh it might be securely reckoned as at least as much if not something more. Still were my present income secure the difference of amount would not be worth the change – but I cannot expect to continue much longer as Examiner, and again, looking still further forward I do not see what is to come of one in London as he approaches the close of his career – you enable a man to retire but we have no means of doing so, and *I* cannot save money.

I confess were I twenty years younger or had I even my London life to begin again I don't know that I should not prefer London, because past experience would suggest a more profitable line of proceeding, & in London resources are within reach of an active man which are denied him in Edinburgh — But in actual circumstances I am disposed to think that Edinburgh would be a better place for me — A man in my position in London ought to lay himself out for Medical appointments, he ought to *pretend* at least to practice & try to secure an Insurance Office or two by which he might have some £500 a year addition without a great deal of trouble. This is what I would do had I to begin again —

I shall say nothing about the kindness of my Edinburgh friends but keep to business & therefore I tell you directly that if you could manage to save me from entering the field as a candidate I should be happy to become Dr. Monro's successor[1] on the terms you mention. But I have a great aversion to being a party in an Edinburgh contested election and I think on consideration of my present position & the duty I owe to University College, my desire to avoid a competition will not be looked on as arrogant. Indeed the Council ought to see that if they desire men already tried in other places, they must smooth the way for them.

I have some other things to ask you about – especially as to the Practical Anatomy Class and as to Goodsir's position.[2] Supposing I had the Chair what is he to do? I should have great objection to a totally independent demonstrator and it is clear that with Goodsir's talent and pursuits he could not long continue as a mere subordinate — Is there any chance of Jameson's[3] multifarious subjects being broke up (into two at least) – when his Chair becomes vacant. Comparative Anatomy and Zoology – for one and Mineralogy, Geology and Physical Geography for the other — Were this done Goodsir would be the man for the first.

The second part of the Anatomy will be out in a few days – the General Anatomy is not yet completed — These things grow on one in the working, and I have been led to treat the subject more fully than I intended – especially *Bone* in which I wished to introduce the results of pretty extensive observations of my own & this could not be done with clearness unless at some length. For the Brain also I am responsible. A young friend took notes of my Lectures on the Brain as a basis, with this he incorporated something more from sources which I indicated & then I finally put his MS into its present shape — There *is nothing new in it*, but I think it contains a fair account of what is known & it is well illustrated with figures. The part would have appeared this week but that the printers wont work in the holidays — I am busy on the sequel —

With best wishes of the season your sincere friend

W Sharpey

[On the back of the envelope:] Let me hear again before you let my resolution be known beyond trustworthy friends.

---

[1] I.e., in the Edinburgh Chair of Anatomy.
[2] John Goodsir had assisted Monro as demonstrator in the teaching of the anatomy course since 1844.
[3] Robert Jameson (1774–1854), Professor of Natural History at Edinburgh.

26

London, 11th February 1846

My dear Syme,

When I definitely communicated to Dr. A. Thomson my views as to removing from London, I had fully considered my position and prospects here and came deliberately to the opinion that it would be right for me to embrace the opportunity of passing the remainder of my days in the scene of my early exertions among attached & old friends and in a most honourable station. In adopting my resolution I believed that I had considered every possible contingency & I felt confident that as I was not activated by light considerations there was not the remotest probability of anything occurring to shake it. It is true that from the confidential nature of our correspondence I was debarred from communicating with those about me but I confidently believed that the announcement of my determination would be met merely with strong remonstrances and these I had fully made up my mind to withstand. These were my sentiments as you well know when I last wrote you and when I received the Lord Provost's letter on Monday last week & I merely meditated a suitable reply to be at once sent off to a communication which held out a fair prospect of my translation to Edinburgh being accomplished in the way most agreeable to me.

It appeared to me right however when matters had arrived at that point to let my Colleagues here know (in Confidence) my intentions. I felt it would be wrong then to keep them in ignorance of it & at their earnest request I agreed to delay giving a final answer for a few days. You were not wrong in attributing this solely to a feeling of what was due to courtesy for I had not the remotest idea that the College *could* if it would do anything to interfere with my resolution. Suggestions of advantage were made to me but they were *such as I had foreseen* and therefore were not such as could warrant a change. There happened to be an ordinary meeting of our Council on Saturday but I positively refused to allow any of these suggested conceptions to be proposed to them, for I abhorred the thought of its being supposed that I should make the advantageous prospects held out to me elsewhere a ground of demands on the resources of the College which I knew it could ill afford & which though granted could in no way alter my resolution which as I said was not taken on light considerations. On Friday morning therefore I told my friends here again that as I had made up my mind I wished to be at rest concerning the matter and I went home with the intention of writing off to the Lord Provost that evening, I then found the enclosed letter[1] waiting me which I send not for the purpose of convincing you of my sincerity but as a testimony to others who do not know me. On Saturday and Sunday I felt at ease in the belief that I had passed the ordeal, for I assure you, it was one to resist such solicitations (especially from Graham) but on Monday I had the Council to encounter & on that day also a prospective arrangement was urged on me which will have the effect of increasing my income by one half with a present improvement in the mean time. This was a proposal I could never have calculated on, at least it never for a moment entered into my calculations & it *alters my situation* here so very materially that I think I ought not to move & I must write the Lord Provost accordingly.[2]

I already feel to the full how much you and Thomson may reproach me — "Why" – you will say – "did you allow us to urge on arrangements up to the very last & then disappoint us". The only reply I can give is that up to the very last I looked forward with satisfaction to the accomplishment of the same object. For whatever may now seen amiss the blame must rest with me, this letter must be your justification. But so distracted did I feel this afternoon that I had almost written to you inclosing two letters to the Provost, one accepting, the other declining, & placing myself at your mercy to give in which you should think right after explaining to you what had occurred to alter my views; but then I felt that I would be using the University of Edinburgh but indifferently well to occupy its Chair in such a mind; it would damp my very best efforts & be a source of distress to me & I sincerely believe also to you for long after.

But after all allowance for the mischief which I might have been answerable for, I would venture to hope that none of any moment is likely to arise. It is quite plain that Dr. Monro could not have gone on another session; that it will be greatly to the advantage of the Univ*ty* that he has resigned & that by resigning at this season a longer time is allowed for eligible candidates to come forward. It is clear moreover that, *whoever* may succeed him nothing could have been more pernicious to the Univ*ty* than binding the successor to any serious extent with retiring allowance, & the Lord Provost in his letter to me expressed his conviction that the Patrons would never consent to it, & I feel quite assured that nothing can tell better for the University *out of Edinburgh* than the recognition of a better principle in regard to retiring allowances. Pray be so kind as to show this letter to Thomson & ask him to excuse me writing another such. I can only further express to you how deeply I have felt the kind friendship of both, and my present sorrow that I should seem to have so ill requited it.

Yours &c
(Signed) W Sharpey

---

[1] See the next letter.

[2] There is no mention of Sharpey in the minutes of the Council meeting of Saturday 7 February. Richard Quain did, however, attend a meeting of the College's Comittee of Management on 11 February, the day that Sharpey wrote to Syme. At this meeting Quain urged that steps be taken to persuade Sharpey to remain in London. On the following day an arrangement was proposed whereby Quain would surrender a portion of the teaching of the Descriptive Anatomy class to Sharpey. The latter attended on the Committee of Management meeting of 12 February and said that he had notified the Lord Provost of Edinburgh of his refusal of the Chair of Anatomy. The Committee guaranteed Sharpey an additional £150 per annum until the new arrangements regarding the anatomy class could be implemented. See: University College London Records Office, University of London Council Minutes, 1843–53, vol. 4, Session of Council, Saturday 7 February 1846; Committee of Management Minutes, vol. 3, Wednesday 11 February 1846: 'Chair of anatomy and Physiology. Mr Quain's communication'; ibid., 12 February 1846: 'Chair of Anatomy'.

27

Friday afternoon. [6 February 1846]
Dear Sharpey,

You will I suppose adhere to your intention of yesterday not to send a definitive answer to Scotland before next [*sic*] & I send this note lest by any chance you should not adhere to that notwithstanding. Be so good then as to omit sending till the time you intended (next week).
Signed R. Quain.

28

To the Lord Provost of Edin^b.

London 11th Feb. 1846

My Lord,

When I wrote y^r Lord^P on Wednesday last I had made up my mind to accept the profess$^{P.}$ of Anat. in the Univ. of Ed. sh$^{d.}$ the Patrons see fit to honour me with an offer of the Chair and if I delayed giving a final answer to that effect, it was solely from a feeling of courtesy which I conceived to be due to my Colleagues & the Council of the Univ. Coll. on such an occasion & assuredly with no expectation that any thing could occur in the interval to alter my resolution.

Since then however arrangements quite unforseen by me have been proposed which very materially alter my position here and which will prevent me from leaving London.

Desiring to express how deeply sensible I am of the consideration with w$^{h.}$ I have been honoured by your Lordship I remain

My Lord your very obed. Serv.

W.S.

29

London 9th March 1846

My dear Thomson,

I am so glad to hear from you again that I cannot refrain from sending you a line or two in reply on the spur of the moment. I can readily grasp the effect that my letter (to Syme) would have on him and you, but I assure you that had you been here and could you have appreciated my feelings at the time & for some time after, you would have been disposed to reply to me in any other rather than an angry tone.

The subject is to me a painful one and I can now only beg of you to accept of the letter I wrote to Syme as equally intended for yourself —

I am glad you have rec$^{d.}$ the second part of the Anatomy. I regret I should have given cause for Dr. Alison's animadversions for I entertain so great an esteem for him in every respect that I should feel much concerned at exposing myself to his criticism in any serious matter.[1] I have long since passed through the stage of life in which one is seriously disturbed by what are called attacks coming from frivolous or contentious men though I am free to confess that like others I am not indifferent to censure from any quarter; but Alison is a man whom I almost venerate on account of his active and *intrepid* benevolence in the cause of the suffering poor, and esteem greatly for his honesty and kindly disposition independently of his merits in science.

Todd spoke to me of your article and promised to send me the proofs but he has not yet done so – it will be a pleasure to me to look it over and offer my suggestions, but I fear I can add nothing or rather I should say I feel assured that nothing from me would add to its excellence.

I have never said anything against the granular structure of bone. All I know is that the lamillae of the soft part are made up of reticular fibres; how the *earth* is connected with them I dont profess to know — I see granules in growing bone clearly enough – sometimes very large as in the leg metatarsus of young salamander larvae – I believe nature can calcify *anything* & sometimes she proceeds with little ceremony as in some diseased ossifications and in encroaching on articular cartilage – which I call a *crude form* of the process – see what I have said at the foot of page CLVIII —

I have been asked by more than one of the Candidates for a Testimonial – but in actual circumstances I have resolved not to interfere in any way even by the expression of an opinion on the merits of the Parties by the usual form of Testimonial — I have no doubt you will get an able and efficient professor —

I have read Steenstrup[2] (some three or four years ago) his observations are so far to the same effect as those of Sars[3] &c – and perhaps susceptible of more interpretations than one in the present state of knowledge. If you would like it I will send you a loan of the book by the Bookseller's parcel — Has not Van Beneden[4] denied the correctness of Goodsir's doctrine respecting the male of the Barnacles? I heard part of a paper of Goodsir's read in the Royal Society a few weeks ago – on the origin of the so called vascular glands – at least the Thymus Thyroid & Supra-renal[5] – I did not quite understand what he had seen but I can scarcely form an opinion till I see the paper or an abstract of it. He seems to adopt Reichert's[6] membrana intermedia – at least he uses the term. I have no confidence in Reichert as an expositor of things observed, nor in his reasonings respecting them; and I hope to learn you will be able to save us in your "Ovum" from an infliction of the Reichertian system.

The universal opinion here is that the Government is very shaky. I would not give much for the purchase of Sir J. Graham's patronage against even old Jeffray's lip – so I don't think anybody can be cock-sure of Glasgow. Were there a turn-about I have no doubt you would cut all out. Lord John [Russell] would give it you at once. If you did not get it I should next like to see Wharton Jones in it – he has worked hard and although he has been unwise in getting into scientific controversies which excite people against him he is a man of great merit.

With sincere regards to your own Mrs Thomson as well as to your father and mother I am

Your sincere friend
WS.

---

[1] These "animadversions" may have concerned an earlier dispute between Sharpey and Alison over the interpretation of some of the latter's views on the spontaneous motion of the blood. See Taylor, op. cit., note 2 above (*Introduction*), p. 146.
[2] Johannes Iapetus Smith Steenstrup (1813–97), Danish biologist.
[3] Michael Sars (1805–69), Norwegian marine biologist.
[4] Pierre Joseph van Beneden (1809–94), Belgian zoologist.
[5] John Goodsir, 'On the supra-renal, thymus, and thyroid bodies', *Phil. Trans. R. Soc.*, 1846, pp. 633–42.
[6] Karl Bogislaus Reichert (1811–83), German embryologist and comparative anatomist.

30

35 Gloucester Crescent, London
23^d Decr 1846
My dear Thomson,

Would you kindly look at the inclosed note and kindly aid the Medical & Chir[urgical]. Society in attaining their object.

It is a most uncommon thing for me to find myself in a position to complain of my friends being behind hand in their correspondence – but so it is at this moment – I have not heard from you or Syme for an age —

I can scarcely think of anything at present worth writing you that you do not know already —

Our classes at the Univ^y College (I mean the Medical) are much the same as last year on the whole – but with 6 or 7 fewer entries. Bartholomew's I am told has 3 or 4 more new ones this year than last. King's much as they were – with no increase. The Borough Schools fallen off.

Have you seen Weber (E.H.)'s Paper on the vestiges of female structure in the male generative organs – and on other topics connected with the reproductive function?[1]

Theile's article on the Anatomy of the Liver in Wagner's Handworterbuch is worth looking into.[2] Amongst other things it contains a curious account of the mucous glands of the biliary ducts. I have been lately examining these funny glands & I find Theile's account quite true – I did not inject the ducts – indeed that is not requisite.

A queer but yet a promising practice has just been tried here of stupefying patients by the inhalation of vapour of Sulphuric Ether in order to render them insensible during surgical operations – Liston cut off a man's thigh in the [University College] Hospital on Monday while the individual was under the influence of Ether – I did not see the operation but I am assured the man felt nothing. [I]t is true that the leg was whipped off quick enough – but the patient had got back to his bed before he felt any pain.[3] — I suspect however that inhalation of Ether vapour if long continued or repeated at short intervals may in indiscreet hands do mischief by causing bronchitis – at least I am aware of this happening with more than one student who has tried it as a substitute for laughing gas.

I believe the Ether is inhaled off water by means of a vessel and pipe resembling a hookah – the water frees it from the sulphuric acid with which it is liable to be contaminated.

The use of ether for the above purpose is I understand a Yankee invention – the suggestor of it a Mr. Bigelow[4] of whose name you must have heard.

Let me know whether the proposed free trade modifications of your Statuta have been adopted[5] – whether they are as yet in operation & what is the promise. I wrote to Syme on the subject a good while ago – expressing my persuasion of the necessity of bringing up the fee to that of the University but also of the great practical difficulty of enforcing the observation of such a condition unless the private teachers agree to appoint a Receiver. Even then I fear *some classes* will not be in a fair position for Grinders[6] will give tickets for Lectures on certain subjects – Materia Medica – Med. Jurisprudence – Pathology or even Physiology, and if they charge full price for these tickets the holders of them will have some compensating abatement or privilege as grinding Pupils – I fear they will be too slippery for you —

Can you suggest any book or books, old or new, for the Sydenham Society? Their edition of Hewson by Gulliver,[7] tho' his annotations refer to matters of detail, you will I think find useful – Gulliver has taken immense pains to have his references correct. As a general rule however I am rather averse to annotating works, especially of older writers. With Hewson it is well enough, for really excepting in the chemical history of the blood, which indeed he does not treat, he is a better guide than many of our own day. But writers a century or two gone by – cannot be made new by means of notes – any notes therefore to such old books should be merely explanatory or occasionally critical – never supplementary – and in any case annotation can only be trusted to a very discreet Editor –

> With best regards to Mrs Allen
> believe me My dear Thomson
> Your sincere friend
> W Sharpey

[1] Ernst Heinrich Weber, 'Zusätze zur Lehre vom Baue und von den Verrichtungen des Geschlechtsorgane', *Arch. Anat. Physiol. wiss. Med.*, 1846, pp. 421–8.

[2] [Friedrich Wilhelm] Theile, 'Leber', in Rudolph Wagner (ed.), *Handwörterbuch der Physiologie mit Rücksicht auf physiologische Pathologie*, 4 vols. in 5, Brunswick, F. Vieweg, 1842–53, vol. 2, pp. 308–62.

[3] This is a singularly offhand account of the first British use of anaesthesia in a surgical operation.

[4] Henry Jacob Bigelow (1818–90), the American surgeon associated with the first use of anaesthesia in an operation at the Massachusetts General Hospital, Boston in 1846.

[5] These remarks seem to refer to a proposal considered by the Senate of Edinburgh University in 1845 that the lectures of extra-mural teachers should count toward graduation in the same way as those of University Professors: see College Minutes, op. cit., note 20.8 above, 1844–55, vol. 4, pp. 65–7. The Town Council accepted this proposal despite the University's opposition. Lengthy litigation followed, which culminated in a decision against the University by the House of Lords in 1854. See: ibid., 1855–61, vol. 1, pp. 429–30 for a summary of the affair.

[6] I.e., the tutors who provided intensive instruction to Edinburgh students prior to examinations.

[7] George Gulliver (ed.), *The works of William Hewson, F.R.S.*, London, Sydenham Society, 1846.

31

My dear Sharpey,

The Transactions of the Royal Society of Edinb. are in the library of the Medical and Chirurgical Socy. up to vol ix part (1821) and we now think of completing the series. The Publishers of the last part we have are W. & C. Tait.

Can the volumes wanting be procured at a reduced price? — Will you be so good as to give any information on the subject, or if you cannot do so from your own knowledge will you oblige me by writing to Dr A. Thomson respecting this matter — We also in the quest of completing the set, would wish to continue to procure the vols. as they come out. May this object be attained by exchanging the Med: & Chirurg. trans. for the Trans. of the Royal Socy. of Edinb. — The answer would depend on the wish of the Royal Socy —

Will you do what is needful as soon as possible. The facts to be ascertained are
1st. The first price of the vols. wanting.
2nd. The price at which we may procure them.
3rd. If the Socy in Edinb. are likely to exchange Trans. on application being made by us —

Yours mo. truly
R. Quain

23 Dec 46 35 Gloucester Crescent, London

32

16th March 1847
Private
My dear Thomson,

I know you are a friend of Sir Wm Hamilton[1] and I think I heard lately that you were attending him professionally.[2] My colleague Mr. De Morgan[3] has been corresponding in a friendly way with him upon some question in Logic, and from a letter from Sir. W^m which I saw today I am sorry to find that the upshot has been that Sir W^m. is seized with the notion that De Morgan has availed himself of disclosures in this correspondence to appropriate to himself certain original views belonging to Sir William. Now from all I know of De Morgan I feel quite assured that Sir W^m must be labouring under a misapprehension. De Morgan feels he must go on and is going on with the publication of what he conceives belongs to him and must take his chance of the consequences, in the mean time he writes Sir W^m to say that in the mean time he abstains from further correspondence by private letters on the subject.[4]

I write you this privately, to inform you that (although Sir W$^{m's}$ letter is really *something beyond* a plain spoken one) De Morgan is activated by no pettish spirit in avoiding private controversy – he feels that in the end it must become public if Sir W^m persists in his persuasion. and that this persuasion on Sir W^{ms} part – erroneous as De Morgan conceives it to be – is no reason why De M. should forego his claim to what he considers his due.

De Morgan spoke with me to day about the matter and I know he is distressed especially at any angry misunderstanding at present knowing that Sir Wm Hamilton has barely recovered from a serious illness — He applied to him for information on certain points in the History of Logic because he knows & believes him to be incomparably the most learned man in that and in kindred subjects not in this country merely but of all men in existence. De Morgan had been long engaged with the "syllogism" before he applied to Sir W^m.

I write to you *privately* to give you an idea of the feelings De Morgan has on the subject and in the hope that if Sir William should possibly speak with you on the matter you may at least restrain Sir W$^{m's}$ naturally impetuous tho' honest disposition from doing any thing hastily and rashly. A cool judicious friend might do much good in this matter.[5]

Letter 32

I had a copy of your Father's Biographical memoir for which I have to thank Dr William [Thomson].[6] I am much pleased with it, but I fear that no memorial of your Father will well convey to them who had not the happiness to know him any adequate idea of the *personal* form and influence of his presence and conversation and oral instruction – above all his unquenchable enthusiasm in favorite pursuit. I can fancy I trace your Sister's hand in some passages — What a piece of meddling fastidiousness in Chambers[7] to go out of his way about the letter to Lord Lauderdale[8] – I should be glad to know how many of those that heard him (of whom I was not one) could have with any face pretended that they had rigidly observed the rule he professed to lay down! — He forgot that Lord Lauderdale was something more to your father than a "lay man" —

Would you be good enough to look at John Hunter's figures of the incubated Egg in Vol 5 of the Catalogue of the Hunterian Museum and explain what you understand by the representation of a layer of amnion reflected over the yolk-sac and another over the allantois – both different from the general continuation of the peripheral part of the serous layer or false amnios.

In Plate 75 Fig. 1 it is marked c and b.

Have you seen Pouchet's recent book & figures on the periodical separation of ova?[9]

Your sincere friend

W Sharpey

[1] William Hamilton (1788–1856), Professor of Logic and Metaphysics at the University of Edinburgh.
[2] This is the only reference in the correspondence to Thomson being engaged in medical practice. Hamilton had in 1844 suffered a paralytic attack from which he never fully recovered.
[3] Augustus de Morgan (1806–71), Professor of Mathematics at University College London.
[4] In November 1846 de Morgan read a tract 'On the structure of the syllogism' before the Cambridge Philosophical Society; he developed the arguments contained in this paper in his *Formal logic*, published in the following year. De Morgan had corresponded with Hamilton about the history of the Aristotelian theory of the syllogism and the latter accused de Morgan of plagiarizing his ideas.
[5] Sharpey and Thomson's efforts at mediation in this dispute are recorded in several subsequent letters in the collection: Sharpey to Thomson, 1 April 1847; Thomson to Sharpey, 5 April 1847 (with a letter from Hamilton); Sharpey to Thomson, 8 April 1847. Only the last of these long letters is reproduced here because, although of some interest to historians of philosophy, they do not bear upon the principal themes of the Sharpey-Thomson correspondence. These letters make it possible to identify Sharpey and Thomson as the anonymous "friends" in Edinburgh and London referred to by both protagonists in the dispute: Augustus de Morgan, *Statement in answer to an assertion made by Sir William Hamilton, Bart., Professor of Logic in the University of Edinburgh*, London, Richard and John E. Turner, 1847, p. 6; William Hamilton, *A letter to Augustus de Morgan Esq. of Trinity College Cambridge, Professor of Mathematics in University College, London, as to his claim to an independent re-discovery of a new principle in the syllogism*, London and Edinburgh, Longman, 1847, pp. 39–40.
[6] William Thomson, *Notice of some of the leading events in the life of the late Dr John Thomson, F.R.S.L. & E., formerly Professor of Surgery to the Royal College of Surgeons, and of Military Surgery, and more recently Professor of General Pathology in the University*, Edinburgh, Stark, 1847. Taken from *Edinb. med. surg. J.*, 1847, **67**: 131–93.
[7] Presumably, the Edinburgh publisher and writer Robert Chambers (1802–71). Apparently a reference to John Thomson's letter to Lauderdale written in 1830 discussing what was to prove the terminal illness of George IV: it was reproduced on pp. 184–5 of William Thomson's life of his father in the *Edinburgh Medical and Surgical Journal*. Chambers apparently questioned the propriety of John Thomson discussing clinical questions with a "lay man".
[8] James Maitland, eighth Earl of Lauderdale (1759–1839), Whig politician and author. He met John Thomson in Edinburgh in the winter of 1799–1800; the two founded a chemistry class, which met in Thomson's house, for gentlemen associated with Parliament House. In 1806, during the short-lived Whig ministry, Lauderdale was instrumental in securing John Thomson's appointment as Professor of Military Surgery at Edinburgh University.
[9] Félix Archimède Pouchet, *Théorie positive d'ovulation spontanée et de la fécondation des mammifères et de l'espèce humaine, basée sur l'observation de toute la série animale*, Paris, J. B. Baillière, 1847.

33

Your Brother W^m is in London, and is to take dinner with me at 6.

London 8th April 1847

My dear Thomson,

I am obliged to write you very hurriedly in reply to your long & perspicacious letter —

I of course know that a man's belief is or ought to be the result of his conviction, & that he cannot alter it according to his will. So long therefore as Sir Wm H. retains his sense of the evidence before him it would be absurd to expect an alteration in his belief — He may consider his allegations as not made and not written – that of course is all he can do – to say his belief was changed while his conviction derived from the evidence before him remains unchanged would be to say what was not the fact. For a like reason I can hardly think Mr De Morgan could be expected to confess himself in error as to the time he wrote down his second view – unless he is convinced of having been in error – and for this reason I dont think he could have compiled with the proposal in Sir W^{ms} letter to you of the 4th which I return inclosed.

There is an inaccuracy in the beginning of that letter which it may be as well for me to correct. — I admitted the possibility of Mr De M. being mistaken in his recollection of the time but I did not say that I thought it "not unlikely".

Again Mr De Morgan was well aware of Sir W^{ms} high position as a metaphysician and would very naturally use expressions of admiration in speaking of it. But he said I did not believe he knew Sir W^m was *Professor of Logic* – this remark had reference to your suggestion that he might have heard thro' some pupil of Sir W^m that he held peculiar views on the syllogism — For all De M. knew to the contrary Sir W^m might still have been Professor of Universal History — His eminence as a metaphysician as you and I well know is independent of his position as professor. —

I wish to give no opinion on the manner in which Mr De M. has conducted his correspondence with Sir Will. – whether judicious – whether in good taste or with a due regard to courtesy – and this I say without any insinuation unfavorable to Mr De M – that is not the question between the parties — Without giving an opinion I will nevertheless remind you that Mr De M. asked Sir W^m to state his views because he *offered to do so* & because they were published sufficiently to secure the right to them.

I fear I shall go astray if I attempt to [. . .] a comparison of Sect. III of the Paper with the addition — It seems to me plain that the addition professes to give a complete theory and states distinctly and in plain words that the author ascribes[?] definite quantity to the subject and predicate of propositions — He then also clearly draws definitely quantitative conclusions — But in section III. he distinctly quantifies the middle term which includes the predicate of something, and he gets a conclusion expressed *within limits as to quantity* which is all he gets in the addition — In Sect. III he expresses by a fraction what is illustrated by which numbers in the addition — But again I must deem[?] that I have no confidence in my own judgment on this point, and, fortunately it is not needed.

Sir W$^{m's}$ "articulate statement" indicating what Sir W$^{m's}$ discovery *would do* led De M. to infer its nature, – it was able to do just what his own would – it was therefore most probably the same – he furnished Sir W^m with some developments of his own view, but as far as I know he even now is not clearly informed of what Sir W$^{m's}$ system is —

De Morgan wrote out his disputed paper at the commencement of the Session when there were few class exercises to look over – the session begins on the 15th October.

Sir W$^{m's}$ letter proposes that De M. should confess himself wrong in asserting that his view was written out before seeing the "requisites" – to admit he may have got a hint and through confusion of thought cannot see that he got it, to correct himself in what he has no reason to believe a mistake as to time, or rather to *assert the contrary of what he believes to be true* – in order that Sir W^m may relieve him of a different and not greater imputation. I wish the affair could be settled without a "paper war" but I see plainly that De Morgan thinks that some sort of imputation of equivocal conduct will be apt to stick to him unless Sir W^m very decidedly & unreservedly clears him, and I think he is much more disposed to bring matters to a public arbitrement, than to settle them by any proceeding which might at any future time leave room for uncorrected misconception or misrepresentation by parties not now in the field. I regret this very much for various reasons – but especially for my respect for both parties. I feel that my voluntarily assumed office is at an end, and come what may I can say that it would *be equally* painful for me if either should suffer in the controversy — Indeed were it not so I think you will give me full credit for saying that from old recollections I should regret more that Sir W. H. should be wasted than his opponent except in so far as the matter touches Mr De Morgan's honour — In any case I am sorry for De Morgan – for, with every allowance, I feel it will appear ungracious in him, young, vigorous and hopeful, to allow himself to fall into a public wrangling with one so much esteemed and deserving as well as obtaining so much sympathy (even from adversaries) on account of his late melancholy affliction — Had it been a dozen years ago I could have looked gaily on the lists.

I wish you would look in to the subject of the development of the spermatozoa. You will have to notice it I presume in your article for Todd – unless there be a special article on the subject – I will gladly send you a loan of Koelliker's book if you desire it.

I have been trying an exper^t of Edwd Weber first mentioned by Volkmann in his article Nerven Physiologie[1] in Wagner's Handwoerterbuch and since then by E. Weber himself in his article Muskelbewegung[2] in the same work. The exp^t was intended to show a different effect obtained from stimulating a nerve and a nerve centre respectively. A current or rather a series of interrupted currents from an Electro-Magnetic Rotatory apparatus is sent through the Sp. Cord of a frog – a tetanic state of the body is produced & this *continues* a little time (a minute or two in my expts) after the application of electr. ceases – but according to Weber & Volkmann when the same state is caused by passing the electricity through a *mere nerve* it ceases *the moment* the electricity is discontinued.

I do not get the same result – I find the tetanic state persists some time after the electricity is stopped in *both cases* – it may endure a *little* longer in the case of stimulating the cord than in that of the nerve but there is no absolute difference. There is no need in using the apparatus with a Natural Magnet — The ordinary Coil &

generating cell used medicinally for giving shocks has just the same effect – indeed I just tried the latter & and find the result diff^r from that of Weber and Volkmann. I then tried the Natural Magnet – but the effect was practically the same. I wish you would repeat the Exp^t for I saw Todd the other day who told me he got the same result as the German Physiologists – I had not then tried it – I used strong large frogs.

Yours affectly

W Sharpey

I should have told you that I saw De Morgan last night & communicated to him Sir W. H's letter of the 4th – which I now return.

[1] Alfred Wilhelm Volkmann, 'Nervenphysiologie', in Wagner, op. cit., note 30.2 above, vol. 1, pp. 563–97.
[2] Eduard Wilhelm Weber, 'Muskelbewegung', ibid., vol. 3, pt. 2, pp. 1–122.

34

London, 20th December 1847

My dear Thomson,

I am well nigh prevented by very shame from facing you even in a Letter. I must acknowledge that you have met with a scurvy acquital at my hands for your kind letter and specimens of the growing cuttlefish as well as for your attention in collecting sheep — (I may as well mention that I should think another dozen or so will be sufficient) —

I send you very rough proofs of the chapter on the Nerves – a little bit remains which I will forward to you as soon as I can dispose of it.

I truly sympathize with you in your deprivation of Syme's services, but your loss is our gain – and I should trust his own[1] — He comes to us in our greatest need and the handsome way in which he has behaved in all the necessary arrangements demands our best efforts to second him both in his capacity of clinical teacher and practising surgeon — The circumstances as you observe are peculiarly favourable to *Syme in particular*.

I have heard from more quarters than one that Miller is seeking the Clinical Chair — In any case you will have a place vacant and I fear your difficulty is enhanced by being obliged to fill it up — Better it remained vacant than that certain candidates should be inflicted on you — I should think your freethinkers[2] professors Simpson & Miller would for their own sakes as members of the University do their best to get the best man notwithstanding differences of Church or State politics.

As to yourself – I presume this event would only seem to fix more firmly your previous determination to the West – unless you take a hint which I threw out in a letter to Syme the other day – as to a Prison Inspectorship which would make London your headquarters – Still I presume that Jeffray – [. . .] tho' he be – will put the Glasgow Chair first in your choice.

One thing I cannot pass over in silence, that is, the regret which I feel that we are obliged to withdraw Syme from among friends and associates who on a trying occasion have given him so disinterested advice — It is a happiness, however, to reflect on their behaviour, although it is alloyed by regret at their deprivation.

How famously Lord John is working the Bishops – and I doubt not he will force the Dean and Chapter of Hereford to make martyrdom of their *consciences* – for in the way they put their case the [*sic*] speak not of the alternative of sacrificing their *livings*.[3] They will doubtless do the deed even against their *consciences* "virtus post nummos" or rather, speaking of sacrifices, "virtutem ante nummos."[4] There will be no *disruption* here. The Candlishes and Cunninghams[5] of Prelacy have too much to lose —

> With best regards to Mrs Allen
> Your sincere friend
> W Sharpey

Dr Allen Thomson

[1] Syme was appointed Professor of Clinical Surgery at University College on 8 January 1848 following Robert Liston's death the previous year. It was Sharpey who conveyed the offer of the post to Syme: see Syme to Sharpey, 14 February 1847, College Correspondence, University College London MSS.

[2] James Miller was a prominent spokesman for the Free Church of Scotland during the Disruption crisis of 1843. Simpson also sided with the Free Church.

[3] Russell had offended the High Church party by appointing the latitudinarian Renn Dickson Hampden as Bishop of Hereford. Thirteen bishops presented a petition of remonstrance to the Prime Minister. The Dean of Hereford was also opposed to the appointment.

[4] "Virtue after lucre" (Horace, *Epistles*, i. i. 53); and "[sacrifice] virtue before lucre".

[5] A reference to Robert Smith Candlish (1806–73) and William Cunningham (1805–61), two prime movers in the Disruption of 1843.

35

London 35 Gloucester Crescent
28th Dec[r] 1847

My dear Thomson,

Whenever you are about to make a new and important step in life I feel that my old sympathies have lost none of their vividness, and it is under their full influence that I say to you you will do quite right in taking the Glasgow Chair [of Anatomy], and would be quite right in so doing even independently of your third reason —

The only person in Scotland who occurs to me as a possible competitor is [John] Reid, but from what he hinted to me in Autumn I judged that his interest had lain with the late Lord Advocate (McNeill).[1] He seemed moreover to consider that not only your interest but your public claims would ensure you the Chair – and from all I saw of his position in St. Andrews it occurred to me that, though not worthy of his merits, yet it was sufficiently comfortable to prevent a man of his easy temperament from making any considerable or troublesome exertions for a better, particularly without a tolerably assured hope of success.[2]

In London the only person likely to cross you is Wharton Jones, and I doubt not he will put all oars into the water to pull into Jeffray's berth. I should think he has not mended his chance by his proceedings last year nor indeed has his demeanour here been ever such as to secure him much kindly interest even with those who are most ready to acknowledge his scientific merit. Who may exert themselves for him in this quarter I do not exactly know – possibly Sir James Clark, but then *his* man again would be Lord John – with whom your claims are paramount.

I think it further not unlikely that interest may be made for Jones from the side of Glasgow – perhaps by applying through the MP's. In case of your success however, which I cannot doubt, another chance is open to Jones, namely, the Physiology Chair in Edinburgh – and I think that would be *his* best game, and best worth even his first efforts.

As I have said – I don't doubt of your success – but let us take care that there is no failure through mismanagement. Your friend James Mylar has access I think to Sir George Grey[3] – but Lord John and Rutherfurd[4] ought and will I suppose settle the affair.

Until Saturday (ie holiday time) direct to me at my house — The old Porter at the College did not forward your letter and I am constrained to write in great haste to secure the post – With kind remembrances and Christmas compliments to Mrs A – I ever am

My dear Thomson
Your sincere friend
W Sharpey

Dr Allen Thomson.

[1] Duncan McNeill (1793–1874), Lord Advocate from 1842 to 1846.
[2] At this date Reid was Professor of Medicine at St Andrews.
[3] George Grey (1799–1882), Home Secretary since 1846.
[4] Andrew, Lord Rutherfurd (1791–1854), Lord Advocate from 1846 to 1851.

36

38 Gloucester Crescent
9th Jan 1848
My dear Thomson,

I write to acknowledge your reply to my letter – but principally just to *remind* you to try and secure the uterus in any cholera case in which abortion may have recently taken place.

I am glad to see by today's paper that the deaths are not increasing – I would hope from this the disease may be about to pass its acme.

The Board of Health here have so precipitately committed themselves on the question of contagion that I cannot look for any unbiased evidence to be collected by their agents. I have lately been pursuing the evidence brought together by Simpson respecting the last epidemic of Cholera. I wish even greater vigilance were employed on this occasion so as to arrive if possible at a settlement of the question.[1]

In the meantime I sincerely sympathize with the inhabitants of your Town and Neighbourhood in their afflicting visitation.

Yours very sincerely

in haste

W Sharpey

Dr Allen Thomson

---

[1] The question of whether cholera was contagious or was propagated by a "miasma" remained contentious throughout the nineteenth century.

37

London 25th Jany 1848

My dear Thomson,

I this morning rec^d your letter but as yet have seen nothing of your book.[1] I was not aware you would be so sharp with it otherwise I would have sent the remaining proof of the Nerves (which I at the time had not beside me) in order to serve as an answer (in some sort at least) to your queries in your last letter. I now regret I was so dilatory but I dare say it was of little moment after all.

I will tell you fairly what I privately think of the book, that is, the whole truth – and I doubt not it is a performance of which the whole truth may be safely told to all – and that my inclination and determination to give it a friendly greeting will be gratified consistently with rigid justice.

I have just got back into writing order again – and hope to get rid in no very long time of the millstone that has been so long about my neck – I will never again engage in composing a systematic treatise —

I thank you for your correction of neurilema, though it came too late. The word ought certainly to have two m's (if we look to the original Greek) but I did not look to this – and in pursuance of my usual custom of following the usual custom wrote neuri*lema* because it is the commonly followed spelling.

I begin to think that old Jeffray's friends are using his name to draw his salary as some Chelsea pensioners have drawn their pensions long after they were dead & buried. Ought not a non efficient but salaried professor to be shown at stated times at Kirk or Market?

We hear rumours here of Jamesons's expected retirement – and of expecting successors – I hope they will give the place to some worthy man – such as Edwd Forbes[2] – at least that it won't be thrown away on old Fleming[3] or Traill,[4] if Traill be madly ambitious of it. The only good of giving it to Traill which I see would be the getting Douglas McLagan[5] into the Med. Jurisp. which would be a useful move I believe.

My neighbour Mrs Potter tells me of a Mr Jamieson of Rutland Square – having died and left something to the University which would especially benefit the Professor of Mathematics – is this true or what is the truth? The Potters will be interested to know on account of their connexion with Kelland.[6]

Yours very sincerely
W Sharpey

Dr A. Thomson

---

[1] Part 1 of Thomson's *Outlines of Physiology, for the use of students*, Edinburgh, Maclachlan and Stewart, 1848.

[2] Edward Forbes (1815–54), Professor of Botany at King's College, London. Forbes succeeded to Jameson's Chair shortly before his own death.

[3] John Fleming (1785–1857), clergyman and naturalist. Formerly Professor of Natural Philosophy at Aberdeen, he was appointed to the Chair of Natural Sciences at the Free Church College of Edinburgh in 1845.

[4] Thomas Stewart Traill (1781–1862), Professor of Forensic Medicine at Edinburgh.

[5] Andrew Douglas Maclagan (1812–1900), an extra-mural lecturer in materia medica. He succeeded Traill as Professor of Medical Jurisprudence in 1862.

[6] Philip Kelland (1808–79), Professor of Mathematics at Edinburgh since 1838.

38

London
Monday [Undated, written between 26 and 31 January 1848]
My dear Thomson,

Confound the old lumberer! On breaking the seal of your note today I expected of a surety that it contained an announcement of his having been gathered to his fathers.

I have neither spoken nor written a word on the matter to any one save Syme to whom you have already communicated your views —

I remember hearing during the Tory regime that an objection to giving two Regius Chairs in the University to two Brothers would be held fatal to your claim[1] – but it was a frivolous objection and without precedent – and the professors are not likely to try in your case the game of the Bishops in Dr. Hampden's and obstruct a Royal mandate.[2]

I suppose that Miller being already a member of the Edinburgh Senatus will be subject to no further test.

The only step which it appears to me you ought to take or at least consider the propriety of taking – is that of letting Lord John know in some way or other that your views are directed towards the Glasgow Chair — He may not know the respective advantages of the position as compared with your present place – which he may very naturally suppose better – and from not being better informed of your wishes may commit himself with the first influential applicant – or his wife may promise her intercession — If therefore Lady John could be made acquainted with your intentions in any allowable way I think it might be prudent to do so – it is I admit a delicate point, but I think it right to say to you what has occurred to me in reflecting on the actual circumstances.

Your sincere friend
W Sharpey

Dr Allen Thomson

---

[1] William Thomson had been Professor of Medicine at Glasgow since 1841.

[2] See note 34.3, above.

39

London, Monday 31st January, 1848

My dear Thomson,

I got your Outlines [of Physiology] on Saturday afternoon & have by this time dispatched the copies to Carpenter and Todd. I have dipped into and turned over the book throughout, and have begun to peruse it from the beginning with care — The result is that without flattering you I have to tell you that I find it greatly to my taste – it is just what I had reason to expect — Short yet wonderfully complete, everything in due proportion, given with great *neatness* and conciseness yet perfectly clear – and above all in a becoming philosophical tone – which is quite refreshing. The only serious fault in it is that it is not ended.

I will read it over with care in due time and freely point out any smaller "spots" that may catch my eye & make any suggestions that may occur for your next Edition – I have no expectation that such opportunities will be frequent and in any case I may say

"Verum ubi plura nitent in carmine,
    (non ego paucis
Offendar maculis, quas aut incuria fudit,
Aut humana parum cavit natura."[1]

But to turn to something else.

If anything else could enhance the gift of your Outlines (*Part 1st*) it would be the news I found waiting me today, of Old Jeffray's extinction – but are you sure it is not another "Cry of Wolf"? George the Third was anxiously impatient for the demise of a certain Madam Schullenberg whom he sorely disliked and when informed of her death he was not satisfied till one of his Pages actually bore testimony to having seen the old lady laid out — Our young friend John ought to have got a sight of the corpse of the Glasgow Patriarch.

I do not know if Sir George Grey has a medical confessor, or if so who is the man – I shall endeavour to learn — As to Sir J. Clark I know he is or was a patron of Wharton Jones – but whether he will keenly interest himself on his account, especially against one who has personal claims on Lord John – I do not know — Clark's influence would lie with Lord John – I am not aware that he attends Sir George Grey. I don't know that Brodie has anybody to promote – from what I have seen I should judge he has not – and besides although a man supported by him might give some trouble I have no idea he will prevail against you – indeed were it generally known among the medical men of London that you were a candidate I think it would in most instances be considered a good reason for declining to bring forward or to use interest for others. I think you had better write Sir Jas. Clark or perhaps rather get Scott[2] to write him (that is to say if you don't think yourself sufficiently intimate) in order that he may be fully appraised of your application – for then he may probably consider that a sufficient excuse for declining interference at least.

I am greatly obliged for the trouble you have taken about the "young lambs" — There will be quite enough – and as you will, like Syme, soon have to "pack up" – perhaps it might be as well just to make a parcel of them and send by Steamer —

I will order Plates of the Arteries – for in the extremely improbably case of your non-translation it will be easy to arrange with the Publishers – they are still some 50 below the 500 copies.

With best wishes
Yours most sincerely
W Sharpey

Dr Allen Thomson

---

[1] "But when the greater part of a composition is resplendent, I shall not find fault with a few blemishes which inattention let slip in a hurry, or to which human nature failed to be alert." (Horace, *Ars poetica*, II. 351–3). I am obliged to Mrs Betty Knott-Sharpe for identifying and translating this quotation.
[2] Possibly John Scott (d. 1853), Physician to the Queen in Scotland.

40

London, 8th February 1848
My dear Thomson,

I received your note last night. I have no doubt your prospect of the Glasgow Chair is all but certain although I learnt last night that Jones is in the field. I saw Sir James Clark and remarked to him that I did not see how your public claims and personal interest could be met by those of any body else. [I]t struck me that he was rather *shy* of the topic especially when I asked him if he knew of other candidates – he said I probably knew that Jones had applied but added that he had very little chance of succeeding.

Lord John is likely in a few days to have another piece of equally important patronage on his hands which it may cost him more trouble to please people in the disposal of, than happens even with a Glasgow Chair — The old Primate though younger than your predecessor (being 82) is not so tough grained – and is said to be dying – I suppose this will keep the Bench right on the Jewish disability question.[1]

Poor Peter (Peebles) [i.e. Handyside] is not to gain his great plea yet, – why does he not stand for Miller's place or Syme's if Miller is not translated?

Yours very sincerely
W Sharpey

Dr Allen Thomson

---

[1] On 11 February 1848, William Howley, Archbishop of Canterbury, died. He had been opposed to the removal of civil disabilities against Jews.

41

London, 14th Feb. 1848

My dear Thomson

I then may congratulate you at last – on coming to your Kingdom – but write you also for another purpose.

When I last saw you you expressed a wish to join the Royal Society – and on mentioning your name to some influential members, they expressed a corresponding wish to see you among them —— A new method of admitting members is now in force and according to the new law the names of all candidates of this year must be given in before the first meeting in March – so that there is no time to be lost. The entrance money is £10 and the annual subscription £4. for which you may compound by a payment of £60 in addition to the £10. in all £70. – but if you have given a paper which is printed in the Ph[ilosophical] Trans[actions]. £20 is deducted from the composition money – so that you have your choice of paying the annual subscription for a year or two in expectation of sending a paper – or of clearing off all at once.

You had better send me a list of your publications with the place and date – for although I know them you can furnish me with a correct list with less trouble than I can make it for myself – I should not think of mentioning in the nomination paper more than two or three or your chief memoirs – but it will be well to remind me of them all.

[William] Baly called on me to ask after your place in Edin. – but he will not move till he learns Reid's ultimatum – I have written to Reid to ask him ——

Syme is well and hearty – he went his round at the Hospital today for the first time – but he will not begin lecturing for a few days —— He has been anxiously looked for by the students who are all glad he is come.

Tell me how you would like your copy of Quain's Arteries – I presume *not folded* – and merely *loose* in a portfolio – so that you can use them singly in Lectures ——

You showed me a plan you had made of the fasciculi of the medulla oblongata – could you favour me with a copy? Or with the original for a week or two that I may have it enlarged as a diagram and then return it.

Yours very sincerely
W Sharpey

Dr Allen Thomson

42

London May 18. 1848

My dear Thomson,

I have been tardy in thanking you for your kind attention in sending me the Embryo sheep – as well as for your long and agreeable letter but events have occurred here of such a disconcerting character that I must postpone for the present more sober matters ——

You will have heard that a Dr. Mr. [*sic*] Cooper[1] raised a storm against us, which together with the malicious comments of the Lancet had the effect of causing some effervescence on the occasion of the public distribution of the prizes[2] — Still the tokens of disapprobation directed against myself and Quain on that day (for there were none against Syme who was well received) were mixed with a larger proportion of manifestos of a favourable kind — The hisses came chiefly from some of the men in Town formerly students but not now attending and the whole affair would I daresay passed over with little notice had Quain held his tongue.

For my part I cared not much about it because I felt confident that it was merely the safety valve and when the malcontents had cut their hiss the whole affair would soon be forgotten and things go on smoothly. But as you know Syme has taken the very inaccountable step of retiring. The grounds he assigns in his letter to the Council are first that he is satisfied he could not undertake the whole or any part of the course of *Systematic* Surgery without endangering the harmony of the School – and 2[dly], that if he did not lend his services in the systematic course and still retained his place in the hospital as Clinical Professor he would embarrass the Council in the necessary arrangements for the surgical department of the Institution[3] — He therefore sent in his resignation.[4]

Now he forwarded his determination on certain assumptions as you see and yet he gave neither the Council nor any one connected with the College any opportunity of explanation for he sent in his resignation and applied to the Lord Advocate for the Edinburgh Chair[5] before he let me know what he was about. Moreover if he embarrassed the Council as he presumed by staying, he embarrassed the Council and us all infinitely more by deserting us. On my endeavouring to reason with him he generally ends by telling me that the step is taken irrevocably and that it is no use to say more.

He declares that he is perfectly pleased with the behaviour of his Colleagues towards him and even still more with that of the Students but he has done what he has done with perfect assurance that it is the right course.

You now know *just as much as I do* in the matter – except that he professes to think the Council wishes to urge on him to undertake the entire course of systematic surgery – now, whatever my opinion of the advantage of Syme's doing so might be, which I strongly stated to him – *before I knew he meditated leaving*, still the Council or rather a Committee of the Council merely desired to know whether in the event of its being deemed expedient he *would be disposed* to undertake the duty. I am sure they never thought of urging it on him — When I asked him his sentiments in the matter about three weeks ago he *seemed pleased* at the *idea* provided a suitable hour should be assigned to him. I wish to pass no opinion at all censuring him for the step he has taken but I lament it both for his own sake and ours.

What I have said between the first and last pair of double lines I would wish you in the mean time to keep quiet, for I fear any cause of irritation that might make matters worse.

Now among the rumours afloat there is one that I am looking after the Physiology Chair in Edinburgh and I beg you will be so kind as give such rumours if it reaches

Edinburgh the most unqualified contradiction. I have never had any such intention and gave no one any reason to suppose so.

I can't help thinking that whether or not it was the best possible thing for Syme to move to London, it is an ill advised step now to turn back – I have every reason to believe that he was getting on most promisingly with practice. — All his Colleagues (except Cooper now no longer one) were pleased with his accession among us and ready to support him. The Students fully appreciated his value – and there was the most agreeable prospect of harmony and prosperity in store for us when all at once Syme makes a spectacle of himself and of us by his present move.

The people I trust will either find it unaccountable or take their own way of accounting for it.

You may congratulate yourself more and more that you have a quiet and comfortable prospect before you – which is secure against the caprice of anybody.

Yours very sincerely
W Sharpey

[1] Samuel Cooper (1780–1848), Professor of Surgery at University College London and Surgeon to its Hospital.

[2] See the *Lancet*, 1848, **i**: 425. In his farewell address to his students, Cooper complained of the conduct of "one or two of his colleagues" who had "exercised undue influence over the Council and the Senate — an influence which permitted them to assume an unjust power in all the proceedings of the College". The *Lancet* took up Cooper's cause and demanded that the vacant chair be advertised.

[3] Robert Liston, Syme's predecessor in the Chair of Clinical Surgery, had assisted Cooper in teaching the systematic part of the course until shortly before his death. Cooper had informed the College Council on 20 November 1847 that he would continue to require such assistance in the future (Council Minutes, 1843–53, vol. 4, 20 November 1847, University College London Record Office). When Syme's appointment was confirmed by the Council in December 1847, it was noted that "it is the desire of the Council to place him [Syme] in the same circumstances as Mr Liston, with respect to other lectures; but as by a previous Minute of Council the whole subject of the Chair of Surgery is to be taken into consideration in March next, no express arrangements can now be settled." (ibid., 18 December 1847).

[4] Syme's resignation was received by the Council on 13 May 1848 and accepted by the Committee of Management on 17 May. (University of London Committee of Management, 1840–51, vol. 3, Wednesday 17 May 1848). The "Cooper-Syme affair" is discussed at length in Taylor, op. cit., note 2 above (*Introduction*), pp. 246–51.

[5] Syme returned to Edinburgh in July 1848 to resume the Chair of Clinical Surgery, and remained in this post until his retirement in 1869.

43

London, 31st Aug. 1848
My dear Thomson

It is perhaps just as well I did not write immediately after seeing the two letters of yours of which I have hitherto been so rude as to take no notice — There were sores open then which are now skinned over, and though you had no hand in inflicting them I could scarcely have avoided offending you with them in my correspondence — All I shall say now is that I am really sincerely happy that Syme's retreat from London has been so soon succeeded by his comfortable reestablishment in Edinburgh and that he is not likely to suffer either in circumstances or professional consideration by his untoward experiment.

I presume you are well on with your arrangements for taking possession of your Kingdom in the West but perhaps before you move I may have an opportunity of seeing you as I start on Saturday for Dundee by the Steamer and mean to make Forfar my headquarters for a fortnight from which I may make a raid across the Forth for a day — I am just writing the last pages of the Anatomy so that like old Jeffray it will be finished at last – I wish I could have had you at my elbow not to nudge me to the work but to aid me with your good counsel on various points – but now it is done and "nescit vox missa reverti"[1] – faults must remain faults.

I hope you are proceeding with your Outlines of Physiology – your retirement from the Chair need not prevent you – it is rather indeed a motive – Your successor[2] has been here, I had a long talk with him, but *entre nous* he seems to have notions of the duties of the Chair which in my opinion are rather suited to his own convenience than the requirements of the University. The [Monthly] Journal [of Medical Science] he tells me will be controlled by two able Editors. I am sorry for them for I am satisfied that the duty of editing a Journal is the most pernicious kind of occupation that a young man of promise can engage in, that is, if he writes in it. All that is gained is a readiness with the pen, desirable I grant, but not at the cost of gaining also the habit of hasty judgment.

I have still to thank you for the foetuses some of which I have given to Marshall for an inquiry respecting the foetal changes in the great veins with which he has been occupied — Quain's arteries arrived safe none the worse for the "pumice stone" of the Edinburgh "Socii".

I suppose Todd will soon be clamouring for "Development", before you come out with it I must have another "pull" at the Decidua now that the millstone of "Quain's Anatomy" is away from my neck —

Regnault[3] the French Chemist is I hear occupied with a great inquiry on Respiration which will I hope furnish accurate data for the future – I am told the apparatus he uses (which was paid for by the government) cost not less than £3000 – it is said to be a masterpiece in its way —

I did not forget your inquiry as to a prosector – but on the whole I should think it will be better for you to take some one who is accustomed to the Scotch ways — You will of course perhaps have to try several before you get one properly to suit, what between conceit and ridiculous expectation, when they are clever or uselessness when stupid, it is no easy matter to make a permanent arrangement.

Poor Wm Macdonald[4] has found his way here in the course of his Pilgrimage – what he will get to do I really cannot see – I suggested popular lectures at some of the Institutions and offered him diagrams – but he is unfortunately one of those for whom his friends can do little. —

With best regards to Mrs Thomson
    believe me

Your sincere friend
W Sharpey

---

[1] "Delere licebit Quod non edideris; nescit vox missa reverti": "the written word unpublished can be destroyed, but the spoken word can never be recalled". Horace, *Ars poetica*, I. 389.

[2] John Hughes Bennett (1812–75), who became Professor of the Institutes of Medicine at Edinburgh in 1848.

[3] Henri Victor Regnault (1810–78).

[4] See note 5.3 above.

44

35 Gloucester Crescent, London
4th January 1849
My dear Thomson

I wish you & yours a happy new year.

Your letter was very agreeable as I longed to hear how you had fared. The account you give of your new flock has entertained me greatly, but I feel assured that with so earnest a teacher they will soon be thoroughly broke in to your own ways. I never had any doubt as to your success in fulfilling the duties of your chair or as to your commanding the respect and engaging the good will of your pupils & I cannot understand on what grounds the men you could mention could have anticipated a different result.

When you kindly accompanied me to the Railway station last September you seemed to be anxious on my account, in consequence of the villanous [*sic*] attempts by the Lancet &c. to stir up the Students here against me — I am glad to say there was not a murmur – I never was better received in my life, and matters have proceeded very smoothly, ever since — Arnott[1] was favourably received and is well liked as a lecturer and teacher — So far all is well – but I daresay you have heard that we have suffered a very considerable diminution in our new entries. Mine will be about 20 or 25 below the average of the two or three last years. Of course you can understand the reason as well as I.

I applied to my Colleague Mr Key[2] to learn whether he could point out any person likely to answer as a Tutor to the little boy you mentioned. He had just then heard from Mr Wilkins a most intelligent Pupil of his own who was at Bonn that young Dr Gesenius a Son of the celebrated Professor of that name[3] was desirous of spending some time in England but felt that his means would scarcely allow him to gratify his wish unless eked out by some literary or educational employment during his stay. At my desire Professor Key wrote to Mr Wilkins to ascertain through him whether such an occupation as you speak of would meet Dr Gesenius's views. Mr W had always written in the highest terms of Dr G's character and qualifications — He speaks of him as altogether a gentleman both in appearance and manner – well educated and accomplished, well acquainted with the English language & with a decided taste for the study of our literature – I think he chose Chaucer's poetry as a subject for one of his academical exercises.

I herewith forward you a letter from Mr Wilkins in answer to Mr Key's inquiry as the most direct way of affording you the information we have obtained. After perusing it pray return it to me. You will see that it is important for Dr G. to know *early* whether his services would be required. To explain his readiness to accept a charge seemingly so insignificant for a scholar, I may mention that I authorized Mr Key to say that the pupil was the son of a person of rank and of high position in this country.[4]

As to my present occupation I have been looking again at the uterus but I believe that there is not much more to be done at this juncture on the points which had previously been engaging my attention. The question of the *genesis* of the placenta I fear does not promise enough in return to make one devote much pain to its solution. I must next look to the condition of the inner surface of the uterus after parturition & endeavour to ascertain with some exactness how much of the altered membrane comes away in the shape of the decidua. But really the day is so short and so *dark* and so *cold*, that I fear I shall do very little till the fine weather sets in. In the mean time I have examined the lining membrane very carefully in several uteri unimpregnated – some which had been so others which never had.

Many thanks for your second part of your book – which I see will be of *great use to me* when I come to the nervous system.

Your sincere friend
W Sharpey

Dr Allen Thomson

[1] James Arnott, by then Professor of Surgery at King's and University Colleges, London.

[2] Thomas Hewitt Key (1799–1875), Professor of Latin at University College London.

[3] Friedrich Wilhelm Gesenius (1825–88), writer on English language and literature, son of the German orientalist and Biblical scholar Friedrich Heinrich Wilhelm Gesenius (1786–1842).

[4] Sharpey's letter of 7 March 1849 makes it clear that this "person" was none other than Lord John Russell. Thomson was still seeking a tutor for Russell's son in 1851 when he wrote to George Combe on the subject: Allen Thomson to George Combe, 18 April 1851, National Library of Scotland, MS 7321, f. 78.

45

35 Gloucester *Crescent* (not Terrace)
7th March, 1849
My dear Thomson

I greatly rejoice in the prospect of seeing you one of these days. I have a spare bedroom at your service and I need not say a thorough welcome. Though our dwelling is almost suburban we have threepenny omnibuses at our command in three directions – every $\frac{1}{4}$ of an hour & some oftener – from nine A.M. till 11 P.M. The termini being within three minutes walk of my door.

I inclose for you a card of the meetings of the Royal Society for the present season, from which you will see that there is a near concurrence of a soiree & ordinary meeting in the beginning of May – which I daresay will suit you, unless you could take the 19th & 21st April. You cannot come amiss to my household whatever time you pitch on – but two or three days warning might enable me to make arrangements (scientific) which might render your visit more advantageous to you.

I can easily understand your hard work this winter, but your men will be broke in by another year – & then, too, your rules and methods will have become almost venerable – a traditional authority as opposed to the right of private judgement, however objectionable in the synod, is the thing needful in the dissecting room – whatever aid Ellis & I can afford towards the compilation of your Rubrick will be cheerfully given — Do you wish me to send you down a copy of the rules as to turning &c. which are followed here? or shall it be when you come up? Mind I don't think that the plan of operations established with us is the best.

I saw Lady John Russell and had some further correspondence with Germany concerning young Gesenius, but after all he preferred travelling for the present & the matter ended – I was sorry, because I fear that Lady John may have lost time by our fruitless inquiry.

Todd is getting on somewhat faster with his *opus viginti annorum* – it is a weary affair & I really grudge giving five shillings for what will bye & bye be got for half price on the bookstalls — Reid's article on Respiration is sensible as might be expected & not too long – but in the last N° there are *nearly two sheets* given up to serous & synovial membranes,[1] an article written by a young man of merit but evidently a youth who has a conceited itch for writing – the easy confident tone of some young hands in the present day is not a little amusing. By the way I wished to look at a point concerning the swimming bladder of fish not long since & had recourse to the recent article Pisces[2] in Todds Cyclopaedia but after very carefully searching it through I could no where find even a surmise that fish possess such an apparatus! I daresay after all you will be in good time for I apprehend there are some heavy articles to come before Z is reached.

Have you seen the new London Medical Journal? Cormack[3] is the unnamed Editor – Dr Williams[4] the chief promoter & Chancellor of the Exchequer (as I am given to understand). I confess my sympathies lie altogether with the previously existing monthly Journal ([John Hughes] Bennett's that was) which has now got a London Editor. Taylor & Walton get some advantages as publishers of the New speculation but have taken none of the risk.

I am sorry to hear of Hutchisons[5] backsliding & misfortune; of course I presume he must be superseded — Would it in that event be infra dig: for the Professor of Midwifery to become director of the Institution? Why should it be so? Unless as to residence I see no incompatibility of the two offices being conjoined in one fit person, who has a taste for both.

Your former colleague Clark[6] is here on his marriage jaunt – I called today but he & his consort were out. Graham says the Lady has great *expectations*. [A]t any rate I hope she is good for Clark's sake for with all his peculiarities I have a great esteem for him.

I think there is a dissertation by some German or other on the anatomy of a stump, but I remember nothing professedly on the subject since the adoption of the existing views concerning the structure of nerves. I have just found the Dissertation I mean — The Title is "De mutationibus, precipice nervorum et vasorum quae in trunco dissecto [. . .]" Auctore C. F. Probst Halle 1832 with 2 plates — On glancing it over I scarce think it will serve your purpose – all is done by coarse anatomy. Perhaps [Otto] Steinrueck De Regeneratione Nervorum Berlin 1838 will be more to the point – for he describes the structure of the swollen ends of divided nerves which have reunited by an intermediate new piece. Still he is more taken up with the new formation & the account he gives of the bulbous ends is not very elaborate. But Steinrueck's is an able performance altogether.

I observe that the learned & worthy Dr Adams[7] of Banchory Ternan has been adventuring on the famous question as to the Hunterian views on the structure of the placenta, and being apparently quite unnerved in the actual examination of such things, has got on the shoals. He argues that nature would never have made the maternal part of the placenta deciduous in man & persistent in the cow – forgetting that no maternal cotyledon nor anything analogous remains in the human uterus after separation of the placenta – & not knowing that we have perfect examples of deciduous maternal vessels in the carnivora. Perhaps you have seen his *judgement* concerning the William Hunterian or Guilmohunterian Preparations, if not, I promise you some amusement by its perusal — The worthy commentator on Paulus could never have injected a densely vascular structure or he never would have reasoned as he does. My friend Lee will I daresay think seriously about reviving his discovery which he has so long let quietly sleep!

Yours very sincerely

W Sharpey

[1] John Reid, 'Respiration', Todd's *Cyclopaedia*, op. cit., note 2.14 above, vol. 4, pt. 1, pp. 325–68; William Brinton, 'Serous and synovial membranes', ibid., pp. 511–41.

[2] T. Rymer Jones, 'Pisces', ibid., vol. 3, pp. 955–1011.

[3] John Rose Cormack (1815–82), an Edinburgh graduate who moved to London in 1847. He edited the *Edinburgh Journal of Medical Science* (1841–7), and *London Journal of Medicine* (1849–52).

[4] Presumably Charles James Williams (1805–89), Professor of Medicine at University College London.

[5] Possibly William Hutcheson, Physician and Superintendent of the Glasgow Royal Asylum. By 1850 he had been replaced by Alexander Mackintosh.

[6] Presumably Thomas Clark (1801–67), former Professor of Chemistry at Marischal College, Aberdeen.

[7] Francis Adams (1796–1861), physician and classical scholar. He published an article 'On the communication between the mother and the foetus in uteri', *Lond. med. Gaz.*, n.s., 1849, **8**: 150–4.

46

London, 9th December 1849

My dear Thomson

I was most glad to hear from you – especially as your tidings were of an agreeable cast. I assure you that I felt anxious on account of my Glasgow friends in the doubt I had as to the possible effect of the Belfast College[1] opening with so many allurements in the shape of exhibitions – in a district from which Glasgow was wont to draw no inconsiderable number of recruits — It was the Gown classes no doubt for which I chiefly feared and they have stood the trial.

Our school is much the same in point of new entries as last year – certainly not better yet still not above two or three under. This is bad enough – but it might have been worse all things considered.

You ask about Morton.[2] There can be no inconsiderable doubt that his self criminating memoranda found after his death give the true solution of his last act. I have long been aware of his addiction to drinking – & before I was aware of it it had become the subject of remark and of reproof by the Authorities of the Queen's Prison of which he was Surgeon – I dont doubt that whilst the consciousness of hopeless slavery to a degrading propensity was at the bottom of his despondency – this despondency so engendered displayed itself in respect to all his relations in life – In his professional position at the Hospital among the rest – tho' that must have been quite recent. He was depressed on account of the way Old [Samuel] Cooper his father in law settled his inheritance – leaving Morton no control over it – a fact by the way that seems to show Cooper's cognizance of Morton's pernicious habit, and say little for his probity in trying to put him into more responsible positions in the College.[3] As to the letter of his friend Mr George, I regard it simply as a well meant effort to secure his (Morton's) reputation. Morton had been dining with him in the evening before he committed the deed. Mr George affirms (and I believe him) that Morton who was then depressed assigned no other cause than his prospects at the hospital – this only proves that, friend as he was to George, he did not give him his inmost confidence. Morton was making a better professional income than most of his contemporaries in the same line of practice – and in such circumstances how a sane man would destroy himself because he despaired of obtaining a certain hospital appointment is inconceivable. But it is but fair to say that I think he must have been deranged – there is an incoherence in much of what he seems to have noted down – which shows an estrangement of reason – and I have since heard that one or more of his family committed suicide and were addicted to drinking. It is all a lie to say that he became more easily [. . .] by drink from disappointment preying on his mind. So long back as August 1847 *before Liston's* [. . .] he was reprimanded by Cap^{tn} Hudson [?] the Governor of the Queen's Bench Prison for drunkenness – before the deputy Governor & assistant Surgeon or apothecary. The last time I saw him which was one day in Summer he had evidently been drinking – and his conversation was abject drivelling – this was early in the afternoon. In short his was a case of suicide from drunkenness. —

I hope you find a spare hour and spare strength to do a little to your article for Todd. I say spare "strength" for I feel as I get older that I have not the same physical strength to get through with work as formerly — Still it is a great matter to keep one's hand in – you have got through with some troublesome generalities – what comes next must be comparatively plain sailing & familiar matter to you — Put it down and be not too fastidious – Give the thing its first shape at any rate and you may add or prune – or remodel parts with comparative ease. It would be great comfort to you to be able to say – "there it is in black and white – requiring some dressing to be sure – but existing in the body & my mind is free[.]"

I hear very favourable reports of the revolution you have made in Glasgow College – for it seems to be little else – I know nobody who could have effected so much in the time. For myself, I doubt not I should have acquitted myself well enough of the mere professorial duties of your position – but I must frankly say I would not have done half of what you have done in the way of *new organization*. Without far more efficient aid than you have had I should have despaired of bringing order and neatness out of such a mess of rubbish – moral and material – as Jeffray left behind him. By the way I have looked at your regulations for dissection. They meet my view better than the plan we here adopt. The only thing I would suggest is to try the practicability of beginning the dissecting with the Perineum and superficial dissection of the face – then proceed as you prescribe – the brain to be preserved in spirits till convenient.

I have got a pamphlet from Syme on Medical Reform.[4] [V]ery sensible – i.e. in pointing out causes of previous failures, but no way helpful with a remedy. Could the government deal with existing interests as Napoleon or Nicholas[5] – the matter would have been settled long ago. Syme forgets that the Lord Advocate is not the Parliament – Mr Andrew Rutherfurd will not forget the difference in a hurry. Now

Syme's plan is well enough if he could *get it* – but it needed neither a ghost nor a Regius Professor of Clinical Surgery to propose it. My plan and [*sic*] be one Examining board saluted if you will from all existing interests – to examine for *license to practice*. [H]olding examinations in England and in Scotland & qualifying for both countries. The license to authorize practice in *all aspects of the profession*. Degrees to be honorary distinctions and to imply not merely professional but general education in larger [. . .]. Colleges of Surgeons & Physicians to make members & fellows in any way they like, but to confer no public privileges. Licences to be registered as well as all [. . .] professional titles, and controlled by any College of Physicians or Surgeons whatever. University regulations granting degrees to be sanctioned by the Crown which I suppose is visitor in most cases.

Who should make their appearance today but John[6] and [. . .] I was glad to hear from the latter that he was admitted at Balliol & only waited for rooms. It seems strange that they should not have rooms ready for their scholars. The two youths promised to come tomorrow for dinner – so I shall hear all about Glasgow. With sincere regards to Mrs. Thomson & your neighbours – believe me My dear Thomson

Your sincere friend

W Sharpey

[1] In November 1849, the Queen's College of Belfast began its first session; it offered courses in medicine from the outset.

[2] Thomas Morton (1813–49), Surgeon to University College Hospital and to the Queen's Bench Prison. He committed suicide on 29 October 1849.

[3] See letter 42. Cooper had tried to secure Morton's appointment as Professor of Surgery at University College upon his own retirement.

[4] James Syme, *Letter to the Lord Advocate of Scotland, on medial reform*, Edinburgh, Sutherland and Knox, 1849.

[5] Napoleon III (1808–73), President of France, and Nicholas I (1796–1855), Tsar of Russia.

[6] Probably John Thomson, William Thomson's son.

47

London, 14th January 1851

My dear Thomson

I have called on Halley and ought to have written you what I learned from him before now – but as he had communicated with Mr John Mylne, the delay is of less moment – Mr Ja^s Mylne has had a more than usually severe attack of his almost habitual complaint – an affection of his breathing which Halley calls "congestive asthma" — His breathing you know is never very free, but now and then he suffers from exacerbations of his complaint. Dr Southey,[1] a colleague in the Lunacy Commission, has seen him professionally along with Halley and coincides in Halley's views as to the nature of the malady as well as its treatment — They seem to look on the action of the heart as more feeble than natural – and there is some irregularity of pulse and irregularity of its force in the two arms – Still he is not considered in a dangerous state. I suspect Halley is quite right in ascribing the occasional returns of the more serious attacks to want of due care on the part of the patient – Mr Mylne when well – at least when in his habitual health – likes to see his friends – and enjoys chat over his tumbler when a friend looks in of an evening. Such little indulgences – of no moment to a strong man – derange his digestion, put him out of health and bring on his arthritic attacks — Again he frequently makes long railway journeys in the discharge of his duty – at hours and in weather unreasonable for an invalid. — Lastly he is rather a self-willed patient. I am glad to say however that according to Halley's account he was getting better of his present attack – and was become more tractable – Dr Southey had expostulated with him and he sees the necessity of using precautions to maintain health when he has it — A very good suggestion has been urged on him – namely that looking to the polycardial effects of exposure in his case – he should [. . .] to arrange with his Co-Commissioners to take only metropolitan duty in Winter and do all his travelling work in Summer – and from what Halley says I dont doubt but that such an arrangement may be effected — Mr Macquorn Rankine[2] called on me the other day – His fathers young ward has however decided on entering the Army, and his guardians have agreed to let him have his way –.

I yesterday received a letter from Nelson – and have this moment got his MS[3] by Mail – of course I have scarcely yet looked into it but from what I *saw* and heard in Edinburgh and especially from your opinion – I feel satisfied as to the importance & value of the *matter* – and, seeing that he has purposely rewritten it as a paper for the Society I have no misgivings as to the *form* — However I will look it through & give you a candid opinion — It occurs to me that as he was your Pupil and imbibed the taste for such researches from your instruction & example, that the duty of communicating it will fall most appropriately & gracefully to you – of course all you will have to do is to allow me to write "communicated by Dr A. T. &c. &c." at the top of it and hand it in to Mr Bell —

I am glad to hear that all your household except yourself are well – and I trust by this time the exception does not need to be made — The grand-nephew & grand-aunt must add greatly to the amenity of your winter evenings – round the hearth of the Great Grandfather – a thought struck me forcibly on reading what you say of Miss Millar[4] – that the personal recollections of the representatives of a past age are well worth noting and preserving – Miss Millar could tell you much of the personal history (if I may so call it) of the University and Town – and it were a pity not to take advantage of her happy, cheerful and communicative disposition – to glean such information.

I am glad that Lawrie[5] is succeeding in the College — I presume the influx of Andersonians are his Andersonian flock upon Andersonian terms – I have little doubt that he will strengthen your staff – but I should scarcely expect any large permanent addition of pupils from the Andersonian[6] source, for unless I am mistaken the Andersonian subsists as a medical School chiefly by its smaller fees – and there will always be an average number of men decided by such a consideration —

Our entries are bad this winter, yet not worse than I expected — Still I feel it hard that whilst I discharge my duty with as much energy and as I trust efficiency as ever, I should suffer from causes which I cannot help.

The remarks on the Edinb[r] Clinical Medicine Chair was a piece of foolish gossip sent from Edin[r] to the Med[l] Times and foolishly inserted by the foolish Editor – as for me I can truly say that it gave me very little disturbance.[7]

I have heard nothing more of Newport's Paper[8] – but if you have reported favourably I have no doubt it will be printed. –

I have just finished the part of my course on digestion – and it is pleasing to contrast the manner in which we are now enabled to handle the subject compared with the confusion that prevailed a few years ago — Have you read the article "Verdauung" in Wagner's Handwörterbuch by Frerichs?[9] To me it seems an able article – containing a good deal of original determination of doubtful points – very clear and methodical in exposition – full and yet not prolix – and not too much loaded with perplexing chemical formulae and disquisitions — The recognition of the real importance of the Saliva in digestion clears up much confusion and reconciles not a few discrepancies which previously caused mistrust. I suspect the clue to an explanation of the digestion in ruminants is to be got from a consideration of the special use of the saliva. Much of the amylum must be converted into soluble sugar during the thorough chewing and insalivation to which the food is subjected, and then what can be imagined better fitted for thoroughly drawing off this soluble matter by absorption than the many plies — The rennet then extracts the major part of the albuminoid matter. The want of acidity in the Paunch is quite in harmony with this view. –

I don't know yet well what to think of the sugar of the Liver — The fact that the liver yields sugar is no doubt true – I tried it last season when I got Bernard's paper[10] & showed the result to the class – but is the sugar *formed* in the Liver? That Sugar should continue to be produced in Carnivora which have been long kept strictly on animal food – is startling at first – still it is not more remarkable than the production of milk containing sugar in a nursing lioness.

I wish I knew more of chemistry!

Have you got Kölliker's Microscopische Anatomie?[11] It shows a vast amount of work.

I fear we shall never see another number of Goodsir's Annals [of Anatomy] — What ought to be done? A record of the kind is wanted. Goodsir has occupied the field, and others were willing to leave it to him as he had taken up the position — but three numbers are due without explanation of the arrest.

I do not see the Edin. Monthly – but I fear the exclusive plan will not answer[12] — A Journal should be something more than the organ of a particular school – it should be catholic in its objects — One fault of the Edinburgh men is, I think, that they hug one another too much – then when they happen to fall out there is the more bitterness – Goodsir is free from that fault, and Simpson is above it himself, though his great merit serves to engender it in others – affording them something to boast of — Our friend Syme is greatly responsible for it in both ways – and Christison from narrowness of view. — As it is I am satisfied the feeling is decidedly prejudicial to progress — Self satisfaction no doubt is a satisfactory feeling to all men – and in all pursuits – and it helps to make men bold in scientific speculation – but these advantages are more than balanced by inevitable one sidedness – and *security*. Of course *you* will not mistake the spirit in which I thus speak of most estimable men and valued friends – with whom I would be happy to make one of a group – and not the less because I flatter myself that my stay in London has very effectually eradicated from me any tendencies of the kind alluded to, if I ever had them. But there seems no chance of Bennett making an opening for the present[13] —

In speaking of Lawrie I might have minded to tell you that our new man Erichsen[14] has turned out a good appointment — He is a good lecturer and a neat operator – with agreeable and conciliatory manners. The students are getting much attached to him — He is young – but the fault speedily mends itself —

I had very nearly forgotten to refer to a letter which I received the other day from the General Secretary of the British Association Mr Phillips[15] (and doubtless you have got one of the same) intimating my appointment as one of a Committee along with Goodsir, Laycock[16] & yourself to report on the recent progress in the Anatomy & Physiology of the Nervous System — It occurs to me that we should divide the "bundle of sticks"[.] I will be glad to take the *histology* of the subject, as Bennett and the Germans phrase it, Laycock might take the physiology – and you & Goodsir divide between you the special anatomy of the nervous centres the Comparative Anatomy and the Development. Each thus undertaking a department of a vast subject, and each in the end having the whole submitted for his approval or suggestions — If we really do undertake so much unpaid labour – this will I think be the best way of getting through it – so that the yoke may be borne with tolerable fairness by each – What thinkest thou?

As the meeting of Parliament approaches we hear the note of preparations for new attempts at medical Bills – notwithstanding the fear of Rome – For myself I have for long ceased to think of such matters – finding that all projects of medical improvement through the legislature have hitherto ended in smoke – I feel that it will be time enough to trouble oneself about them when there seems a prospect of serious action.

Mrs Colvill and Mary are happy to hear of you – and beg to send their kind remembrances.

And recommending myself with sincere regards to Mrs Allen and Miss Millar as well as your brother William and all his fireside, not forgetting my young friend Miss Helen — I ever remain My dear Thomson

Your affec^t Friend
W Sharpey

Dr Allen Thomson

---

[1] Henry Herbert Southey (1783–1865), an Edinburgh graduate who had been a Commissioner in Lunacy since 1833.

[2] William Macquorn Rankin (1820–72), civil engineer and natural philosopher.

[3] Presumably Henry Nelson, 'The reproduction of the Ascaris Mystax', *Phil. Trans. R. Soc.*, 1852, pp. 563–94.

[4] One of the sisters of Thomson's mother, Margaret Millar. Their father John Millar was prominent in Scottish political and cultural circles in the late eighteenth century.

[5] James Adair Lawrie (1801–59), Professor of Surgery at Glasgow University since 1850.

[6] The Andersonian College, founded in 1796, was a rival in Glasgow to the University. Its medical school remained independent until the twentieth century.

[7] A reference to a report of 28 December 1850 that Sharpey was to return to Edinburgh as Professor of the Institutes of Medicine after John Hughes Bennett had been translated to a new Chair of Clinical Medicine. See *Medical Times*, n.s., 1850, **1**: 678.

[8] Probably George Newport, 'On the impregnation of the ovum in the Amphibia', *Phil. Trans. R. Soc.*, 1851, pp. 139–42.

[9] F. Th. Frerichs, 'Die Verdauung', Wagner's *Handwörterbuch*, op. cit., note 30.2 above, vol. 3, pt. 1, pp. 658–871.

[10] Claude Bernard, 'De l'origine du sucre dans l'économie animale', *Arch. gén. Méd.*, 1848, **18**: 303–19. Bernard's claims concerning the glycogenic function of the liver were at first controversial.

[11] Albrecht von Kölliker, *Mikroskopische Anatomie; oder, Gewebelehre des Menschen*, 2 vols., Leipzig, W. Engelmann, 1850–4.

[12] There was, presumably, some suggestion that articles to the *Edinburgh Monthly Journal of Medical Science* should be provided exclusively by members of the Edinburgh school; I have been unable to find any reference to this proposal in the journal itself.

[13] This remark shows that Sharpey was still considering a return to Edinburgh — on this occasion as Professor of the Institutes of Medicine. See Duns, op. cit., note 51 above, (*Introduction*), p. 360, which reveals that Simpson and some of the other Edinburgh Professors were keen for Bennett to take over the Practice Chair "in order that we might get Dr. Sharpey down by this arrangement to our University".

[14] John Eric Erichsen (1818–96) was appointed Professor of Surgery at University College London in 1850.

[15] John Phillips (1800–74), geologist and one of the founders of the British Association for the Advancement of Science.

[16] Thomas Laycock (1812–76), lecturer on clinical medicine and writer on the physiology of the nervous system. He became the Professor of the Practice of Medicine at Edinburgh in 1855.

48

Private
College, Glasgow
19th July 1852
My dear Sharpey,

You have heard of course of Old Thomas Thomson's death and I daresay may through Graham know something of the canvass. From living in the country and seeing very few people I know nothing almost of what is going on.

Immediately after the death some of my medical Colleagues who are attending my Lectures on the Textures (instead of resigning as they threatened on my appointment) placed before me for signature a very high testimonial for Dr. R. D. T.[1] already signed by the whole six, Macfarlane's[2] name not being there as he is not yet inducted. I demurred on the ground first that I never interfered with appointments in the University to which I belonged, and when pressed to sign it because of his having already been in the University and having established a claim; I gave them my real and decided opinion that he was not the man to confer reputation on the University or to teach the class efficiently. I then learnt not much to my surprise for I knew the fact already that their opinion was exactly the same as mine viz. that personally he was a very estimable man *but* &c. &c. and that this was only to keep out any *worse* man. Who they pointed at I do not know but the principle among my Colleagues & in Glasgow generally of "Our own fish guts &c." is remarkably strong and is heightened by a strong Edinophobia. I think it is with some the fear of breaking in upon the nice snug way of granting degrees, and with others a general prejudice.

I saw how disagreeable a position I was to be placed in and at first agreed to sign a moderate testimonial merely stating that we were satisfied with the manner in which the duties of the class had been conducted, which was all we as professors had to do with: but even that I knew was a little strong and so on reflection as I saw I had no right to dictate terms to the others I declined simply to sign on the first grounds of noninterference which I had stated to them. At the same time stating my personal regard for Dr. R. D. T.

A great push has been making for him. The old tory way of making appointments through the Duke of Montrose[3] has been revived since the accession of the tory ministry, and I believe the old Principal who has renewed his youth on the occasion has been seeing visions of the entire recovery of the ascendancy of the Church and tory party in the College. The last appointment of Macfarlane was made entirely through him. We have yet to see what is to turn out of that appointment. He never was famous as a Lecturer and only gave a few lectures on Clinical Surgery long ago and he is not understood to have been reading much since that time.

The Principal is reported to have said that if Mr. [Thomas] Graham was not in the field, he would recommend Dr. R. D. T. You may believe I feel very anxious in regard to this appointment. If a good man is placed in this chair there is some chance of our medical school rising out of its present state, and the University will derive eclat from the new professor in so important a position as a good man will occupy in Glasgow and in the scientific world generally: but if we are to have nothing better than heretofore, adieu to anything like a revival of the school or the system of teaching, and all reflected credit on the University from the occupant of the chair. I need not conceal from you that what with William [Thomson]'s death and the appointment of his successor, if the Chemistry chair were to undergo no improvement I should feel myself quite alone in the University and would have scarcely any tie to it but the salary paid to me.

I do not know in the least what are Graham's emoluments from his office or from other sources in London: and he must know pretty well what those are connected with the chair in Glasgow. The salary is £250: The fees are at present, I suppose about as much or more, and might be at least doubled; and I understand the income is doubled by references and Analysis which of course in good hands would also increase. It is no doubt the most important Chemistry Chair in Scotland if not in the country, and if Graham thinks it his interest to apply for it or only to signify that he would accept of it, there can be no doubt he would get it and would make a capital thing out of it. But you must have gone over all this with him already, and I am anxious to know what has been passing with you on the subject.

Thomas Anderson[4] and your old pupil Blyth[5] are candidates, and they are both excellent men. Free kirkers &c. are excluded as the tests are inforced. I have no idea who more may be in the field – in fact I am very ignorant of what is doing: but I argue rather favourably from the delay that seems to be taking place, and conclude that the claims of candidates must be under consideration. The principal has been seemingly ill too for a few days which may have assisted in putting off any immediate appointment if it was to depend on him.

I have a deal to say to you. When may we expect to see you down? You will have your room either at Greenhall with Mrs William or at Millhaugh where Mrs Allen and I sleep, on account of Miss Millar who is now becoming somewhat infirm.

I am hard at work with development for Todd. He has printed two sheets and I have advised him to bring out a short number and give me a little more time for the remainder. I wish it were out of my hands[.] I shall feel a great load off me. It is like a constant *night and day* mare on me.

I need not tell you that all my grumblings as above are only for your most private ear, so I beg you will burn them as soon as read, and let me have your news as soon as possible.

I have a little *private* story about a memoir of William & the Edin. Monthly journal which will interest you when I have time. It is sent to the Medical Gazette and Times & will appear shortly.

Ever my dear Sharpey
Your sincere friend
Allen Thomson

Remember we count on your coming down and staying as long as you can to gather mushrooms and health, and to leave behind you consolations and contentment among the grumbles.

---

[1] Robert Dundas Thomson (1810–64), nephew of Thomas Thomson. He assisted his uncle in the teaching of the chemistry class from 1848.

[2] John Macfarlane (1796–1869) succeeded William Thomson in the Chair of Medicine at Glasgow.

[3] Both the third and the fourth Dukes of Montrose held the office of Chancellor of Glasgow University. The former had been instrumental in Thomas Thomson's appointment to the Chair of Chemistry in 1818. The fourth Duke, James Graham (1799–1874), was a staunch Tory.

[4] Thomas Anderson (1819–74), a lecturer in the Edinburgh extramural school, eventually succeeded in the Glasgow Chemistry Chair.

[5] John Blyth, Professor of Chemistry and Practical Chemistry at Cork College.

49

London 20th July 1852
My dear Thomson,

Your agreeable letter has just come to hand and finds me much in the humour for writing you a reply – for indeed I had been actually meditating on the expediency of sending you a missive for the last two or three days. It is true I have little to say of myself – but let that little be first said. Having long felt the difficulty of mastering (I mean rightly understanding & *retaining*) the results of modern physiological chemistry, I have e'en begun again at the beginning and worked for the last three months as a humble practical pupil in the Birkbeck Laboratory, deeming practice (always accompanied with an inward ratiocination) the best means of attaining my end — I thought it best to go through the entire of *in*organic analysis before tackling the Animal Chemistry and am already tolerably expert in qualitative analysis. I shall continue the work as I can find time which I hope to do for 2 or 3 hours a day during the busy season. I have been greatly pleased with the occupation which I find a most salutary mental exercise. My only fear is forgetting it all too easily again. Another object I had in view was to judge for myself how far practical Chemistry (studied in a Laboratory) can [. . .] be recommended to the better class of Medical Students.

Now on the strength of my chemical proficiency I am not going to offer for the Glasgow chair, just yet, but in consequence of my pursuit I have been thrown somewhat more than usual among those who are capable of forming an opinion of the candidate and are not slow to speak it out – and the first remark I will make is that I was greatly pleased and indeed *proud* to think that my friend Allen Thomson had not lent himself to, but, exercising the moral courage he possesses, had abstained from the very lamentable proceeding of his medical colleagues in regard to the vacancy. Were I Walpole I would discourage the system of introducing sons & nephews as assistants in the hope that a convenient opportunity would be found for fixing them on the Institution – positively I would prefer *an equally good* candidate free from this objection – and I assure you I have heard various independent persons express a similar opinion —

Graham is quite a settled denizen of London – he is much more in London society than I am – he is a man of independent property – and as to consulting chemical practice he might have as much as he likes — For some years back he has except in particular cases (Government Commissions and a few others) declined that kind of work – and gives himself almost entirely up to scientific inquiries — The business of assaying in gold & silver as a control on the Mint and for the Bank of England he manages through a highly intelligent assistant who is quite up to the routine — This business itself has vastly increased of late so that he will need some additional help.

I mention this to you to show that the prospect of *practising* as a consulting Chemist would not take him to Glasgow – and although attached to his native town, he keenly relishes London life and the Scientific Society here – moreover he is much devoted to the University College, so that he is out of the question.

Here we look on Anderson as the probable winning horse, and most likely to reflect credit on the Glasgow University by his scientific eminence. Blyth is also well thought of especially *as a teacher* – Sir R. Kane[1] was here the other day – I did not see him – but he spoke to Graham in the highest terms of Blyth and of his services in the Cork College. Williamson[2] seems to look on Anderson as having better claim than Blyth on the score of original working. But, out of all the names mentioned I have not heard a word in favour of *R. D. T.* Of course, I need not ask you to keep my remarks quiet – as I have no mind to meddle in the matter or to commit other people by repeating their opinions.

The Principal is a sworn friend of Graham's uncle and Minister of Killearn – and I understand he has answered applications for his interest saying that he will support Graham. Now perhaps this is to get rid on easy terms of importunity – or it may be earnest – but in any case he can hardly be committed yet to R. D. Why should not Anderson not apply to him and refer him to Graham for an opinion?

I hope Todd will not comply with your request — Depend on it it is better to be forced to get through with a task which has become irksome.

It seems that the case of Lizars versus Syme is really to come into Court – about the end of the month — Will the verdict – whichever way it be – alter the position of the parties in publick estimation?[3] Quarrelling is "a fault" but to carry professional quarrels into a court of law is worse – it is "a blunder" – or as I would rather say it is sheer folly; and I am glad *in this case* that Syme is not the pursuer. Syme would meet with more concurrence but for the spirit in which the reviewing especially of surgical works is conducted in the Edin. Monthly. But I have often said before the great mistake of most of the Edinburgh men is that of self complacency. You have all need of each other – all, the best of us much to learn – and the worst may not be utterly incapable of helping the best on an occasion.

I am glad to say that after an unpardonable delay Nelson's Paper is being printed in the Phil Trans. – I will tell you more of it when we meet — I have now beside me a *lengthy* MS. by Newport[4] – In his former researches he could never observe the spermatozoa any deeper than the surface of the enveloping jelly of the ovum. One day that he was showing the fact to me I noticed two or three spermatozoa at a considerable depth within the jelly – tho' not in the interior of the ovum, soon after Mr T. Bell and Mr Busk[5] who had gone like me to see his expts. noticed abundance of the creatures (or plants) deep in the jelly – and since then Newport has seen it to be common, & has sent a supplementary communication to the RS. corrective of his former Paper — He is a most expert dissector and manipulator of minute objects – but my faith in him generally as an observer begins I confess to waver – I never cared much for his attempts at higher generalization in Physiology – I must however say that excepting in that point referred to I found all the matter he showed to us to be as he had described them.

I will gladly avail myself of your invitation again to enjoy a retreat to Clydeside with your good converse & companionship. You know I am tied here till the Examinations at the Univ: of Lond. are over – that is till after the third week in August —

With kind remembrances to Mrs Allen Mrs William & Miss Millar

Believe me my dear Thomson

Ever your sincere friend

W Sharpey

[1] Robert John Kane (1809–90), Professor of Chemistry at Apothecaries' Hall, Dublin from 1831 to 1845, and subsequently Professor of Natural Philosophy to the Royal Dublin Society.

[2] Alexander William Williamson (1824–1904), Professor of Practical Chemistry at University College London.

[3] In 1853 the long-running feud between Syme and John Lizars (1787?–1860) again led to litigation. On this occasion it was Lizars who claimed that he had been defamed. See John A. Shepherd, *Simpson and Syme of Edinburgh*, Edinburgh and London, E. and S. Livingstone, 1969, pp. 126–7.

[4] George Newport (1803–54), London surgeon and naturalist. His paper 'On the impregnation of the ovum in the amphibia' appeared in the *Proc. R. Soc. Lond.*, 1850–4, **6**: 214–7.

[5] Thomas Bell, one of the Secretaries of the Royal Society; and George Busk, a Fellow of the Royal Society.

50

Wrote 26th [This is in Thomson's hand.]
35 Gloucester Crescent, London
19th May 1854
My dear Thomson

Under an uneasy feeling that I owed you a long letter with information on various matters which I had promised – I have been actually deterred from any *intensive* communication for nearly a twelve month!

Leaving yet a little while that more full communication contemplated — I am induced to write on the present occasion by Dr. Tweedie who asked of me to forward you the inclosed documents respecting Dr. Kirkes[1] candidature for the Assist[n] Physicianship to St. Bartholomew's Hospital. Baly, his Senior is unopposed, there being two vacancies. Kirke's opponent is the son of one of the Old Physicians of the Establishment; Dr. [Clement] Hue and carries with him all the influence among the Governors which nepotism (rarely pure & undefiled by personal claim) can command. The Hospital Medical & Surgical Staff are to a man supporters of Kirkes – and they as will others of his friends are endeavouring to bring public opinion to bear against the close [*sic*] system – Dr. K has accordingly applied to various medical authorities of weight, for testimony in his favour and has been already very successful in obtaining it. It is not considered of much use to have such testimony from men in *London* connected with Hospital or Educational Establishments (I mean beyond that already given) inasmuch as the lay governors will merely look on such as evincing as esprit de Corps of the *Medical* bodies in London as opposed to lay patronage — In reality it is a manifestation against the dominancy of *personal favour*.

Inclosed is a printed document which has been pretty extensively signed by medical men *out of London* and Dr. Kirkes is very desirous of obtaining your testimony to the same effect — As it is adapted for Hospital officers – perhaps the best way will be either to make such alterations by erasure &c. as will suit it for you – or to write out a certificate of the same import – mutatis mutandis – This will save you the botheration of *compounding* a testimonial, which is to me very sickening work —

Your nephew John is here at present. He has been twice at the R[oyal]. S[ociety]. and has been introduced to the President's Soirees where he has met with several leading Engineers and men of science much interested in his present undertaking. He spent a couple of days with Admiral Smyth[2] down at Aylesbury – who told me he is greatly pleased with him.

We are approaching to a close with our Session at the R. Society & I have got on very pleasantly in my post. You will have received the first two numbers of the Proceedings in their new dress, and should you have any brief communication suited for them I will be happy to receive it. We desire to give early & brisk circulation to all that is sent to the Society – and in particular we wish it to be known that an author may send a scientific communication – to be read to the Society & published in the Proceedings although he may not think it of sufficient importance to appear in the Philosophical Transactions — Or should he in the progress of his inquiry fall in with results deserving of being made known in the mean time they will be received & published without any prejudice to an embodiment and more full exposition of them in a more elaborate memoir which he may subsequently present for the Transactions.

Yesterday I fell in with Mr Charles Henry at the R. S. apartments. He inquired after you and after Williams family with great interest; and expressed himself in very grateful terms respecting the attention and kindness he had received at 80 George Street.

Are you likely to be here soon?

With kind regards to Mrs Thomson
   believe me always
      my dear Thomson
         your sincere friend
           W Sharpey

Dr Allen Thomson

P.S. You had best send the document for Dr Kirkes to his own address.
  2 Lower Seymour Street Portman Square
This will save time.
          WS

Our Medical School has *rather* improved this winter, there having been a small increase (8) in our *new* entries.

[1] William Senhouse Kirkes (1823–64), Demonstrator in Morbid Anatomy at St Bartholomew's Hospital since 1848. In 1854 he defeated John William Hue in a contest for the post of Assistant Physician to the hospital.
[2] William Henry Smyth (1788–1865), admiral and scientific author. He was especially interested in astronomy.

51

Wrote 17th [This is in Thomson's hand.]
London, 15th August 1855
My dear Thomson

I should have written before now to say that, unless I have been forestalled by more distinguished candidates, I mean to be your guest during the Meeting of the British Association [for the Advancement of Science].

I shall suit the time of my arrival to Mrs Thomson's convenience entirely – for I daresay her arrangements will require no little management – You know I dont care how I am stowed away.

To finish with my own plans — I have been harassed all Summer & incapacitated for work by a irritation of "boils & blains" which have perseveringly kept up a continuous warfare against me – one healing and another coming on – like the reinforcements in the Crimea – I have tried various prescriptions but without any marked benefit. In these circumstances I think it will be best for me to get away to the north as far as Dee Side in the first instance – so soon as I can get away from here – and then go to Syme a few days before the Glasgow work – or else go to you before & to him after it is over – but I think it will be best to go first to him. All this however as far as Glasgow is concerned is subject to your arrangements – I mention this particularly on the present occasion because of the turn about in your household which the *Sorners* [i.e., sojourners] of the Association will necessitate.

Next as to a more important Guest. I have a letter from Kölliker from Würzburg – in which he informs me he will come to the Glasgow meeting. He proposes if my arrangements suit to travel from London with me to the meeting. From what I have already said you will see that I must be out of London long before that time. But the point is to have quarters for him in Glasgow — Now I dont know whether you will or can take him in – but I presume there will be no difficulty in having him hospitably received in some private house, and I would be glad if you would arrange this and write to *him* to that effect. He leaves Würzburg today or tomorrow, will be a fortnight in Paris & will arrive in London ab^t the 1st Sep^t — A letter will reach him in Paris addressed "chez M. Dumont. 6 Rue Vivienne." I will write at once to Kölliker to say that you will inform him where he will be received in Glasgow. —

I am just packing up to go to another residence — I have been 11 years where I am & wish to get nearer again to the town & to my work – accordingly I move to 33 Woburn Place – where after Friday letters must be directed. The House is larger than I want & yet I could not find one with such accommodation as I wished unless a large one — I intend to take a Pupil to reside should I find a proper one – so as to take off part of the rent – and I hope to have more elbow room — Moreover we shall always have quarters for you & Mrs Thomson whenever you like to honour London with your presence —

I have just heard that Alison has resigned – and that Bennett is likely to try for his place — Should he be successful of which I have great doubts – the chair of the Institutes would be vacant & this might require me to put a question to myself – What do you suppose might be reckoned on as the income from the Chair & Graduation Fees? Bennett is here & I was asked by Tweedie to meet him to day at dinner as he wishes particularly to see me – But as I wish particularly not to see him, that is not to involve myself by any talk on the matter, which might afterwards give rise to misunderstandings & require explanations – I have excused myself.

Bennett made a great mistake in choosing Edinburgh for the scene of his activity – His merits are of a kind which would have told much more with the London Public – I mean of course in *practice*.

Your niece Helen Thomson called here the other day – Mary was unluckily down at Woolwich – & since then has been so taken up with getting our house in order that she has been unable to call. I hope we shall repair this delay however before Helen leaves London.

Yours very sincerely
W Sharpey

P.S. I need hardly say I do not wish to seem to be actively interested about the Edinburgh changes – for after all it will be time enough to think about the matter when Bennett moves – if that ever happens. Should Christison wish the place I daresay he would get it before Bennett and I think there will be candidates from here — You will see that the question I put to you is done *quietly* in order to be furnished with a reliable statement of the return of the Class.

What effect has the new regulation (admitting a proportion of extra academical lectures to qualify for the Degree) had on the Edinb. College Classes?[1]

                W.S.

[1] See note 30.5 above.

52

Ballater, Deeside
2[d] September 1855
My dear Thomson

Your letter addressed to me at London (of the 27th aug.) reached me yesterday after making the circuit of London & Edinburgh. I came to Edinburgh on Monday evening by the express train – Staid at Millbank all Tuesday – saw nobody but the Symes came to Arbroath on Monday and here on Thursday night. Here I found my Sister M[rs] Martin of Dundee & her husband – in Lodgings. I am in the Hotel. Syme must have me back in Edinburgh to dinner on Wednesday – so that I must leave this Tuesday morning. I will come to you on Wednesday the 12th or sooner if I can get away. I will let you know the time beforehand.

You allude to the ongoings in Edinburgh — I suspect Bennett is too unpalatable with the Profession there to succeed, with all his backing, I guess too that he is over urgent and not over discreet. I shall be satisfied either way. I made up my mind that if the opportunity of moving to Edinburgh offered itself I would not let it go past – but should there be no opening I am quite contented. In the mean time I have told Syme that I wish to remain quite passive. I have no desire to press myself upon the choice of the Patrons – Still less to mix myself up with Bennett's plans. I thought it right to state plainly that my services were at the command of the Patrons if they desire them, but beyond this I do nothing – and whatever happens I shall be sorry if "ill blood" should be engendered between estimable people on my account.

I think I have been improving for the two days I have been here – I have had a tolerably long walk one day & have twice ascended rather respectable sized hills without in any degree overworking myself. Indeed I think I have a fair prospect of getting set up again for the winter – if people would mercifully and considerately allow a poor invalid to dispose of himself in his own way. As to the [British] Association I am right glad our head quarters are to be at Greenhall and I shall beg to be left at liberty to go & come to the scientific fair as I find agreeable and convenient. Above all I beg you will not confer on me any office which would require attendance on a Section – as I am quite unfit for confinement or for exciting work – indeed I shall rather than subject myself to that, cut the concern altogether & enjoy peace & quiet.

I saw Retzius[1] in London. He & his Prosector are to set out for Dublin – thence to go to Belfast & finally to Glasgow — I doubt if he had got your letter – I mean a private Letter from you – for he had received a Public Invitation – Carpenter promised to write you about him.

Yours very sincerely
W Sharpey

Dr Allen Thomson

[1] Anders Adolf Retzius (1796–1860), Swedish anatomist and anthropologist.

53

33 Woburn Place, London
15th October, 1855
My dear Thomson

I begin to fear that if I do not write this afternoon (even in a hurry) there will be another week's delay in letting you hear from me.

And first as to the Drawings[1] — Your Letter was addressed to *Edinburgh* & did not reach me till Friday morning — I wrote Mrs W^m that day by return of post, decidedly advising the disposal of the collection in the way suggested and with an abatement in the price originally approved by myself – in consideration of their present intended & possibly final destination you will likely see the Letter so I need say no more of it.

After leaving you I spent the remainder of September at Broughty Ferry except the last four or five days – as I came up by Sea & could not cut so close as by rail. Before leaving Scotland I was in good measure prepared for Bennett's defeat – but I confess not for Laycock's success — True, I had been assured on fair authority that Simpson had taken his side – but really I could not allow myself to think that this was the real case, without a supposition as to Simpson's character for fair dealing, that to me was inadmissible. It was quite open to our little friend to further the cause of Laycock or any other Candidate he preferred, but this I considered to be quite incompatible with the language he used & the professions he made.

You will believe me and be pleased when I assure you that I heard of the result with perfect equanimity. An opportunity had offered (in a certain contingency) of my moving back to Edinburgh and Edinb^h friends, it was necessary I should decide what I should do & make known my decision — I concluded that the opportunity should not be allowed to pass & patiently waited the result, not *anxiously desiring* the change nor with sanguine thought as to the event & consequently the event caused no mortifying disappointment.

M[ary] has had a pleasant ten days with Kölliker as an inmate of the house – only I wish we had had *one* quiet dinner by ourselves at home — He left us yesterday Morning – thoroughly pleased with his visit to Britain & full of kind remembrances to yourself – Mrs Allen – the young Ladies & all other northern friends — He read to us Mrs Allen's Letter which greatly gratified him — The approach of Post time and the end of the paper warns me it is time to say how completely I am

Your sincere Friend
W Sharpey

[1] Presumably William Thomson's collection of teaching illustrations.

54

33 Woburn Place, London
27th Oct. 1855
My dear Thomson

I wish you would help me to find as soon as possible *four* more *Acting* Assistant Surgeons for the Turkish contingent.[1] The pay of and appointment is 10/6 a day, and if the candidate has a Diploma & gives satisfaction in the service – he will have a good chance of promotion to a higher rank. He will have a free passage out & home, lodgings or lodging money, free rations except when on board ship. He must agree to serve a year if required and will be guaranteed a year's pay & should his services be dispensed with after a year's duty he will receive a gratuity of half a years pay —

In case of being disabled by wounds or accidents received in the execution of duty – he will be guaranteed a pension.

We had a nice letter from Kölliker yesterday — He found all his people well —

It is rather curious that our Medical School (at Univ. College) should have made a considerable rise this year I have Sixty five new men which is more than I have had since the year of Liston's death (1847). The *average* of new entries for some years back has been about 50 — It will mean an addition of about £100 to my emolument & I may glean a straggler or 2 before the end of the Session —

My enemies [i.e., his boils] have been very vexatious since I saw you. They have now retreated in great part to my back & some nights have made it difficult for me to find a position to rest in. Generally, however my health keeps good — I have sometimes thought that my present trouble is sent to punish me for not enough valuing the robust health I have up to this time enjoyed — They interfere not a little with every kind of exertion both bodily and mental.

With sincere regards to Mrs Allen –

I am ever My dear Thomson
Your sincere friend
W Sharpey
Dr Thomson

[1] I.e., to act as military surgeons in Turkey during the Crimean War.

55

33 Woburn Place, London
5th December, 1855
My dear Thomson

I am glad to accept your proposal for a weekly interchange of letters – as we can neither receive nor dispatch letters on Sunday – it would seem that my day for writing should be before or after the *Sabbath* as you call it in the North – but that is little matter – & perhaps my plan is to write on Sunday & add a word if necessary on Monday before posting.

In the meantime I do not wait for that day, meaning this letter to stand for the first in the series. But now it occurs to me (on the moment) that we should so arrange that one should write after receiving the letter of the other – say that I write on Saturday or Friday and you on Sunday.

I am greatly pleased indeed to hear of your augmentation. I daresay the war has something to do with it – although the improvement is not general in London as far as I can hear. Guy's is said to be good this year, but the reports of most other schools is [*sic*] different.

We had our Anniversary at the R[oyal]. S[ociety]. on Friday – things went off pleasantly notwithstanding a speech from (the eternal) Mr. Babbage[1] not blaming the Council but lamenting or affecting to lament that we had not given the Copley Medal to Mr. Schentz the Constructor of the Surdish Calculating Machine. His allusions to his own achievement in this way were in a very moderate *tone*. But it was curious to see how the consciousness of his own merit peeped out – in short it was one word for Schentz and two for himself. We had a very pleasant dinner afterwards at the Free Masons Tavern at which Sabine[2] & I ingeniously contrived to make Graham propose the President's health, by threatening in case of refusal to give "Her Majesty's Government and the Master of the Mint".[3] We had a table a stupendous Jack (or Pike) three foot & a half long sent by Mr. Whitbread the Brewer who is a fair Amateur Astronomer & is getting a noble Lens ground for his new Telescope — I told him afterwards that some friends (Paget[4] & Burke) who sat near me regretted it was not made into a Skeleton, but indeed we left nothing but the bones.

I have just received Bennett's Introductory Lecture for the past Session[5] – on glancing through it *very rapidly* I was rather pleased on the whole. Will the great Lay-Cock print his first *crow*? I quite agree with you that people must suspend their judgment as to his success, till better assured one way or the other – and with our friend Syme indifference would take the shape of absolute failure — An intelligent student of ours now a resident in the Edin. Infirmary, who came here the other day to take his Degree of M.D. spoke as if the reports of the new man inclined to be unfavourable – that he laid little stress on Auscultation or Microscopic research. My young friend had been Clinical Assistant to Christison & to Bennett.

Having given you the above sample of gossiping – such gossiping as may make the staple of my correspondence, turn now to a little point of business lest I should have to *entamer* a third sheet. I wish to know confidentially *what kind of person* you think Dr. Harley[6] who was once one of your Pupils – and I think an Assistant to Simpson – &

who has since been studying under Bernard, Robin & Verdril – Kölliker, Scherer & Virchow? He has applied to be teacher of Histology & Practical Physiology in our College – and as the Assistant Curatorship of the Museum happens to be vacant he is now filling *that* post *temporarily*.

The case of the Mylnes if not uncommon – two members of a family cut off within a short time of each other – 'Tis almost a wonder James Mylne did not fall before now – considering his infirm health and the way he guided it. I had agreeable news of your Sister through Miss Graham – who on her way to Rome passed a short time at Nice & is loud in praise of Mrs. Mylne's attention and Kindness.

The little packet was from Retzius – containing injections of biliary ducts, which he wishes me to show to Kiernan.[7]

With kind regards to Mrs. T. & family – Your sincere friend

W Sharpey

[1] Charles Babbage (1792–1871), mathematician and inventor of an early form of computer.
[2] Edward Sabine (1788–1883), soldier and natural philosopher, was himself elected President of the Royal Society in 1861.
[3] Graham was appointed Master of the Mint in 1855.
[4] James Paget (1814–99), Assistant Surgeon at St Bartholomew's Hospital, had been a Fellow of the Royal Society since 1843.
[5] John Hughes Bennett, *The present state of the theory and practice of medicine*, Edinburgh, Sutherland and Knox, 1855.
[6] George Harley (1829–96) graduated MD at Edinburgh in 1850. He secured the demonstratorship in histology to which Sharpey refers.
[7] Francis Kiernan (1800–74), examiner in anatomy at the University of London. He was the author of *The anatomy and physiology of the liver*, London, R. Taylor, 1833.

56

Boils
College, Glasgow
9th Dec./55   Addressed to 33 Woburn Place
My dear Sharpey,

I was very much gratified by receiving so promptly a letter from you, and I cannot do less than reply on this appointed day, though I cannot expect to give my epistles that amount of interest which I may count upon from those coming from the metropolis.

It is very annoying indeed that you are still troubled with the boils. What do you say to a trial of the panacea Cod liver oil.[1] I have very great confidence in its changing just such conditions of the system as that under which you are suffering.

Begin with small doses such as a tea spoonful at a time, and I believe a little dry oat meal after it is the best thing to put away the taste and oily feeling in the mouth.

It was indeed a shame to eat such a magnificent pike as you consumed to the backbone. I am in expectation of some fishes from the Mediterranean, through Verany at Nice with whom John Mylne and my sister are acquainted. There are some Torpedoes, Chimeras and Young Sharks from the Uterus, I hope with external gills.

My friend has not come back yet from Western Africa, but he writes that he has been very unsuccessful, having only procured one chimpanzee. I have got from a pupil who was in the North last summer specimens of the Clio borealis, and another interesting Pteropode the Limacina which is coiled, and has a small rudimentary shell on one side. I was interested by obtaining just at the same time a specimen of the only Pteropode found in the fossil state, and that not very common, viz. Conularia quadrisulcata from the limestone shales between the upper and lower coal beds of this district.

We have just had a visit of John Mylne and I daresay you may see him ere long on his passage through London on his return to Nice. He tells me that his brother James has left his family very wealthy. He has given me as a rememberance of his brother William a very nice copy of the reduced edition (Roy. 8vo) of Audubon's American Birds.[2] It seems to exhaust the subject completely and also is a most elegant drawing room book.

We had an interesting meeting or soiree of the Athenaeum on friday evening at which there were some good addresses. The speech of James Moncreiff[3] the Lord Advocate pleased me extremely. His theme was the education of the people and his sentiments were of the most liberal kind. Notwithstanding his being trammelled somewhat by the free kirk[.] I think I can see that he has had enough of kirk difficulties in connection with the education question and that he will be disposed to act more independently when next he brings the subject before parliament.

My number has closed at 170 in all. 155 in one class & 135 in the other. I am introducing more examination than I used to do for I become more & more convinced of its importance in elementary instruction.

I am actually bringing ovum to a close. I expect to have it out of hand by the New Year, and I hope I will not run my neck into such a noose again. After all it will be a failure as I have missed the mark: but I shall be so happy to be done with it that I shall scarcely mind its success.

There has fallen into my hands a beautiful book by Rusconi which I had not seen before viz. On the development of the Land Salamander published posthumously[.][4] The plates are exquisite. There is also a Memoir on the fecundation (artificial) of fishes and the development of fishes published like the other in 1854 by Moyanti and a biographical account of Rusconi by Serafino Biffi. They were sent to the Brit. Ass. and the local Committee paid £6-6 for the carriage; but I mean to have them for our library or myself, so you need say nothing about them, in case any body having less right to them than myself should ask for them.

I am rather annoyed to find that I cannot from my recollection at this moment give a distinct account of Dr. Harley about whom you inquire. In 1847-8 I had a George Harley a *good* student but I cannot recover his face or his qualifications, the more so that there was also a student of the name of Hartley the same year. If I saw him or had any circumstances to awaken my reminiscences I might be able to say more.

But I must conclude I send you a note about my philos. trans. w^h I am ashamed to have neglected so long.

Ever Dear Sharpey most sincerely yours

Allen Thomson

[1] Probably an allusion to John Hughes Bennett, *Treatise on the oleum jecoris aselli, or cod liver oil, as a therapeutic agent in certain forms of gout, rheumatism, and scrofula*, Edinburgh, Maclachlan Stewart, 1848.

[2] John James Audubon, *The birds of America, from drawings made in the United States and their territories*, New York, J. J. Audubon, 1840–4.

[3] James Moncreiff (1811–95) succeeded Andrew Rutherfurd as Lord Advocate in 1851. In February 1854 he introduced an Education Bill for Scotland, which was rejected in Parliament. A second Bill in 1855 was passed by the Commons, but thrown out by the Lords.

[4] Mauro Rusconi, *Histoire naturelle, développement et métamorphose de la salamandre terrestre*, Pavia, Bizzoni, 1854.

57

Wrote 29th [This is in Thomson's hand]
33 Woburn Place, London (W.C)
14th March 1857
My dear Thomson

I heard of Johnny's having had Scarlet Fever after he had passed through the danger, but I was not aware that he caught measles afterwards – It is well he is rid of both. I have a horror when I hear of epidemic scarlatina – not on my own acccount indeed for I am proof against fevers of all kinds – but I know of no disease as suddenly & widely diastrous when it gets into particular families.

I am glad to hear such good accounts of your *class* and of your new modifications of your Lectures – No doubt the more the Student is made an *active* rather than a passive participator in a course of instruction so much the better. I heard good accounts of your Edin. Lectures[1] – and there is much new matter as to the eye which I will be glad to learn from you – when we meet. Since I saw you I have gone somewhat into the question as to the nonstriated muscular tissue – & I am satisfied Ellis[2] must give up the point. The "weeny ovum" is very welcome —

But my main object in writing you at this moment – and my reason for not entering into scientific affairs – is to remind you of your intention & promise to pay us a visit at the end of the session – in company with Mrs Allen. Try and finish as early as you can so that you may have longer time to be here – and as we shall be delighted if you bring Johnny with you – Let you and your worthy spouse sit down and settle at once the time of your coming & let us know & if there is anything I can do to further any purpose you may wish to fulfil in London besides friendship – I will be most happy to set about it —

I am truly sorry to hear your account of Ninian [Hill] – but it is consolatory to think that complaints like his are not now viewed so despairingly by medical men as formerly — How happy to be able to choose Nice where of course he would be kindly tended.

Political people, as you know, are at present in a hubbub – Finsbury is one of the largest constituencies in the kingdom if not the largest & what an atom I feel myself to be in such a mob — Time was when I should have gone to hear the candidates – but now I care little for these things – and have little faith in public men – as such. Every body cries out in admiration of Lord Palmerston's good luck – whatever may be thought of the China business[3] — He must have dissolved at the end of the

Session – with some objectionable points for his opponents to harp on at the hustings – but the Derbyites &c. have stepped in to send him to the country on a question in which he will doubtless be popular – i.e. backing up his agents and boldly taking responsibility. No doubt the China affair is an unlucky business & God knows when it will be settled – but to disavow the proceedings of our People then would not have mended things. Your young Laird (of Hatton) is a Candidate for the Fellowship of the RS. this year – Sir Roderick[4] speaks most favourably of him – indeed every one I have heard who knows him — Remembering your skill in arranging our old house – how happy I should have been had we had the benefit of your aid in our arrangements in Burlington House.[5] We are getting the books fast in – and busy rearranging them – but we shall not be able to use the new hall for our meetings for some time yet.

With sincere regards to Mrs Allen

Believe me

My dear Thomson

Yours most sincerely

W Sharpey

Dr Allen Thomson

[1] Possibly a reference to Allen Thomson, 'On the phenomena and mechanism of the focal adjustment of the eye to distinct vision at different distances', *Glasg. med. J.*, 1858, **5**: 50–69.

[2] Ellis advanced a heterodox theory of the structure of involuntary muscular tissue, arguing that it was fundamentally similar to that of voluntary muscle. See G. Viner Ellis, 'Researches into the nature of the involuntary muscular fibre', *Proc. R. Soc. Lond.*, 1856–7, **8**: 212–3; *idem*, 'An account of the arrangement of the muscular substance in the urinary and certain of the generative organs of the human body', *Med.-Chirurg. Soc. Trans.*, 1856, **39**: 327–38.

[3] In March 1857 a general election was called after the government was defeated on a parliamentary motion calling for an investigation into the outbreak of war with China.

[4] Roderick Impey Murchison (1792–1871), geologist, had been a Fellow of the Royal Society since 1826.

[5] Burlington House was the new premises of the Royal Society.

58

10th July 1857

> By the bye. I have never received any of my photographs. Should I write directly to the *Artists* did you order the six which I wrote to you about?

Hatton House, Ratho

My dear Sharpey,

As I have the expectation of seeing you soon I will not enter upon a long letter, but only write to make inquiry as to the time you expect to be free and to come down to see us. The weather has been a little changeable, but on the whole fine and occasionally very much so, and you will be quite charmed with our beauties if you come in the proper season. I trust therefore to your holding to your intention of getting down in July to see things when they are at their best. I may add that both the ducks & green peas are in a prime state.

I have not been very well for the last week having had a bilious attack brought on I believe by some indiscretions in diet conjoined with the irregularity w^h is apt to be produced by my frequent journeys to Glasgow; but I am getting out of it and hope to be quite equal to walks with you.

My course has done pretty well, at least I have great reason to be flattered with the attendance having had upwards of 100 students and occasionally as many as five professors. It has not given myself great satisfaction, the nervous system as a whole being far too extensive a subject to study in three months and that study being continually interrupted by the necessity of giving expositions of what is known and of providing illustrations.

Tuesday and thursday are my regular days of going to Glasgow: and I sometimes go on a Monday when required, but if I choose I can spend from friday to tuesday quietly here — I shall lecture for 2 weeks more. Now write me a note as soon as you can & tell me your plans.

I am quite out of the Medical world so I am far behind as to what is doing or to be attempted by the Universities in the Medical Bill[1] being in Committee. I always feared that the influence of the Corporations w^d. be sufficient to carry their measure.

I have a printed Circular in reference to a proposal to modify the division of the University faculties especially that connected with "Science" as they call it by which I presume they mean Natural or physical science. I quite approve of some such modification as that pointed at in the Circular. viz. that physical science should have two main divisions. A physico-chemical and another comprehending Natural history & Biology for w^h a convenient name would be desirable. If you have time would you give me a hint whether it is proper for me to reply & give my opinion and what are your views.

I must not delay longer – but hoping to hear from you soon I am ever sincerely yours
Allen Thomson

---

[1] A reference to the Bill which became the Medical Reform Act of 1858. This measure made provision for the registration of all qualified practitioners and created a General Medical Council on which the universities and medical corporations were represented.

59

33 Woburn Place, London
15 November 1857
My dear Thomson

I ought to have replied to your letter before this but something came in the way at the time to make me defer it.

The piece of microscope was left by Kölliker (as he did not go home by Paris) to be taken charge of by Ninian. Mary carefully put it out in order that it might not be forgotten but the Servant, thinking it was something out of its place, put it back into the drawer & so it was left behind. I have already spoken to one or two people who have much intercourse with Paris, to let me know when an opportunity offers – and although none has occurred as yet, still I think we shall more readily find the means of sending it from here than you are likely to meet with at Glasgow – and I shall not neglect it.

Roper has not turned up yet. I find some of the English Geologists speak rather harshly of him – but his reputation is such – with Scientific men in general, that there would be no question as to his being one of the *fifteen* selected for the RS.

I am glad to hear that the "Laird" is to make his appearance at Burlington House on Thursday. I will not neglect to indoctrinate him properly as to Hatton – if the opportunity should offer.

I am sorry indeed for the hard fate of bona fide mercantile Houses in Glasgow as elsewhere – but I find the common reproach of trading on other people's money is especially applicable to your neighbours. I am concerned to hear of the loss which the Mylnes will sustain – and I am sorry to say that Dr. Arrott who is much less able to bear loss – has one or two shares in that abominable Western Bank.[1] I warned him years ago not to trust to these people – but a "wilful man" as they say &c. —

Dr. Livingstone[2] told me that he had once given a supply of the poison used by the Bushmen in South Africa for their arrows – to Dr. Buchanan[3] – Would you ask Dr. B. for some of it? for Koelliker & a little for me. Koelliker has been experimenting with the Upas Antiar. & Tiliste (Tchetick) & proposes to send a communication to the R. S. on the subject — I have been making one or two trials (rather in a desultory way) with Mr. Antiar. Brodie, long ago & found that it stops the heart as Koelliker finds at [. . .] — I find that it extinguishes the irritability of the muscles generally – Koelliker who I suppose has worked out the phenomena thoroughly, writes me the same thing – but it appears to me that the motor nerves lose their excitability also — I have been studying Koelliker's Memoir on Poisons[4] – which is quite a masterpiece – & I have repeated some of his expts. with the Woorrara[?] – very satisfactorily. Koelliker got some of the Antiar & the Tchetick from old Dr. Horsfield who was long in Java & wrote a memoir on these poisons more than forty nearer fifty years ago[5] — I have since called on the old man and obtained a further supply of Antiar for Koelliker (and myself) and I will be happy to give you a little if you should want it. It was examined chemically long ago by Pelletier & Caventou – & more recently by Muller,[6] but I hope Koelliker will find some competent man at Wurzburg to give it a further scrutiny.

My Class is not so good as in 1855 – but rather better (in new Entries – "Gulpins as ye may call 'em" *Lizars*.) than last year. Aitken called today – his book[7] will be out in a few days – I presume that Watson's Lectures will keep their ground but there is nevertheless quite room for another.

Mary sends her kindest regards to you – Mrs. Allen & Johnny. Heartily joining therein I remain

Yours very sincerely
W. Sharpey

Dr. Allen Thomson

---

[1] A Glasgow-based bank which went into liquidation.
[2] David Livingstone (1813–73), medical missionary, I presume.

[3] Probably Andrew Buchanan (1798–1882), Professor of the Institutes of Medicine at Glasgow University.

[4] Presumably Albrecht von Kölliker, 'Observations on the poison of the Upas Antiar', *Proc. R. Soc. Lond.*, 1857–9, **9**: 72–6.

[5] Thomas Horsfield, 'On the Oopas or poison-tree of Java', *Ann. Philosophy*, 1817, **9**: 202–14.

[6] Joseph Pelletier and J. B. Caventou, 'Examen chimique des Upas', *Annls Chim. Phys.*, 1824, **26**: 44–63. I have not traced the Müller article.

[7] William Aitken, *Handbook of the science and practice of medicine*, London, Griffin, 1858.

60

Wrote 31st Jan[y.] [This is in Thomson's hand.]
33 Woburn Place, London W.C
21st Nov. 1857
My dear Thomson

We had a very good opening Meeting at Burlington House on Thursday evening. Your young Laird was there & we were introduced to each other by Sir Roderick, who stood father to him on his admission. We had a very short talk together partly of Hatton — He is satisfied that it yet wants much clearing & he intends paying a visit himself bye & bye & setting about clearing, seriously and on a plan. He would be much improved I think by clearing away his moustache which is broad short & scrubby – something like the brushwood at the west end of Hatton House.

I and Mary are sincerely happy to hear so good accounts of the travellers. It was a very pleasant time for us – the two days they were here – & I need hardly say that we look forward to a repetition of the pleasure when they return. — I was anxious about their success in a lodging – & am right glad to hear that they have got all they want without paying what in Scotland we call a heavy "ransom".

I wish the benefit of your contriving skill, to advise me how best to make a plan for keeping & rearing rabbits. They are dearer than dogs in London and the high price really stands in the way of employing them for experiments. There is a large piece of waste garden ground at the College pretty well out of sight, so that a rabbit hutch would be no eyesore – Make me a plan of one & I will laud you. I wish also to try and keep frogs. We can easily direct a streamlet from our water pipe into a tank for the purpose – How should a tank be made? What would be requisite to prevent the creatures from escaping? Had we a proper place I should try and breed the rana esculenta which is much larger than our frog – It would not be difficult to get a stock from the continent – I have suggested to Lister that his Father's place at Upton would be well suited for such a trial. But I don't see why we might not make it answer at the College.

Yours ever sinc[y.]
W. Sharpey

Dr. Allen Thomson

61

Hatton House, Ratho
10th Aug$^{t.}$ 1858
My dear Sharpey,

I was very glad to hear something of you though indirectly through our travellers. I daresay you will have enjoyed your visit to Koelliker much.

I dined with the Symes yesterday and they like us are desirous to know your Autumn movements and are looking forward to the pleasure of seeing you. I was greatly shocked as you may believe with poor Mr. McVicar's appearance. I saw him only for a minute reclining on a sofa in the upper parlour pale & emaciated and with all the look of a person dying rapidly from Organic disease. I did not get any statement from Syme of the nature of his disease, but I believe it is some tumour either of the stomach or liver. His appetite seems to be nearly gone and I do not think he can live long. There can be little doubt that his misfortunes have hastened very much if they have not altogether caused this sad change. What a comfort it must be to Mrs. McVicar to have all the kindness and attention they receive at Millbank.

Your absence from London prevented me from writing to you on a matter on which I would now wish your advice and if possible your assistance. The second son of my late Colleague Dr. John Couper (bearing the same name with his father)[1] has been a very promising pupil of our school, and having been left some fortune & not having any liking for Medical practice, would fain set up somewhere as an Anatomical or physiological teacher. I have explained to him that a thorough acquaintance with the drudgery of the dissecting room is the first thing he must have to make himself an anatomical teacher, and he is quite prepared to go through it. He would prefer greatly to be in London and I have promised to inquire for him through [you] or Mr. Ellis whether it might be possible for him to get a position in your school as an under demonstrator or something of this kind which would introduce him to the practical knowledge of Anatomical teaching. He is a very sensible quiet laborious young man a very good dissector, has been on the continent for some time, has taken his degree in Glasgow, and is now anxious to set himself up in some employment and means to proceed to London about the 18th or 20th of this month when his engagement as Clerk in the infirmary comes to a close. I will give him an introduction to you and I shall take it kind if you have any thing encouraging to say if you will write me here before the 16th when I expect [him?] to come to Hatton to pay us a short visit, and pray let us know when we may expect you down.

I have a great deal to say to you about University & Medical Bills and other matters, but I dare not venture to begin them at present. All here join in kindest regards and I am ever

Most sincerely yours
Allen Thomson

[1] John Couper, graduated MD at Glasgow in 1858, was sometime Professor of Physiology and Anatomy at the London Hospital Medical School, and later Lecturer on Surgery there.

62

University of London Examin$^{n.}$
12th August 1858 – 3 p.m.
My dear Thomson,

Here I am in a very very hot & close afternoon looking after a set of men at an Examination. I foresee that the looking over their papers will last until Wednesday night or thereabouts.

I have spoken to Ellis about young Couper. It is unlucky that Ellis has already arranged not only with a gentleman to be his regular Demonstrator but also for another to be a sort of Supernumerary. I do not think Mr. Couper would have much or any difficulty in obtaining an appointment at some of the Smaller Schools – but then I don't think that would answer his purpose. A Lectureship even in a small School would (were he ready for it) give him the opportunity of trying his hand at the work & furnishing evidence of capability when something better should turn up – but a Demonstratorship would be of little use — Should he come to London, I will give him a note to some of the Anatomical Teachers in the principal Schools & mention his object – but I presume that their arrangements are already made for next Winter. After all, unless he has some other objective in coming to London now – I think he would prepare himself to be a teacher of Practical Anatomy much better by undertaking to work under you at Glasgow this winter as ass$^{t.}$ Demonstrator. I am satisfied that the work is much better and more methodically conducted at your own School than it is any where in London — This I say as a matter of business & without any nonsensical flattery – and I verily believe Mr. Couper's best plan would be to "bury" himself (as your Father once said to me) in your dissecting room & work constantly and very hard at pure anatomy — I did this in Berlin in the winter of 1828–29 – and I may say that from daylight till 12 p.m. *every day* Sundays – even Christmas day not excepted I carried the work on. It is the only way to learn thoroughly the whole matter of descriptive anatomy – & with you he would have the advantage of learning methodically the art of supervision of Students — I may add that I think it is a defect of several modern Physiologists that they are not sufficiently grounded in rough anatomy – It is a most valuable foundation & specially necessary for the study of comparative anatomy —

I scarcely remember an event which has caused me more concern & excited more sympathy than what has happened to McVicar – I do believe that his loss of fortune distressed him quite as much on account of others as of himself – & I doubt not that had his health not given way he might have soon reconciled himself to his altered circumstances, which would still have allowed him to live at ease tho' not in affluence – but from what you tell me & what I have heard from Lister who is with his friends at Upton just now – I fear he must be rapidly approaching his end. Few will be more lamented.

I hope to see you before October – probably about the end of this month – but I should like to know whether one time would be as convenient for you as another for my visit to Hatton – I doubt if it would be suitable for me to go *to stay* at Millbank while there is so much cause of affliction there – I will write you again shortly —

I have a good deal to tell you about our visit to Germany – which was really very agreeable – I was very *loth* to move from home – but once the inertia was overcome – I was delighted with the motion – my only regret is that we missed the Roman travellers as they passed thro' London —

I see I have filled two sheets – I will therefore reserve the remainder (as the Ministers say) for another occasion —

Yours very sincerely

W. Sharpey

63

Private
The College, Glasgow
26th November 1859
My dear Sharpey,

It is I must say a great grievance that the appointments to scientific chairs in the Universities should be influenced in the manner which it appears is being done by the Glasgow Members in the case of the Surgery Chair.[1]

I understand that through Dr. A. Buchanan's friendship for Dr. Lawrie, his brother Walter Buchanan[2] espouses warmly George McLeod's[3] interests, and that through private friendship with old Dr. McLeod[4] (who is his neighbour in the country) Mr. Dalgleish has with equal warmth taken up the Cause of his son; and that thus these two sapient radicals have taken up the cause of the Candidate of the greatest & most unscrupulous tory connection in Glasgow on the grounds of private friendship and the flimsy & absurd view that Surgeons of Glasgow growth should alone obtain places in its University "Our own fish guts &c."

I really trust that Sir C. Lewis[5] will see through this kind of thing, and I am sure that if he consults such men as Sir James Clark, Sir Ben[n] Brodie and the like, he will have the grounds on which such an appointment is to rest placed in a very different light. I would not wonder but that it may be represented that the University professors are chiefly in favour of George McLeod: but this is entirely false. Indeed when poor Lawrie proposed him as his substitute, we were very averse to take him, partly because some thought him by no means the best person, and partly, because with Dr. Lawrie's own concurrence he had been secretly engaged in a canvass for the chair as if it had been vacant: and Dr. Lawrie's Colleagues went the length of offering among them to conduct the course for him. Dr. Lawrie declined this offer and pressed Dr. McLeod on the Senate, who could then do nothing else but sanction his appointment as substitute: but in order to protect ourselves from any appearance of expressing preference we put upon the Minutes the following: "The Senate in deference to the wishes of Dr. Lawrie accede to his request and hereby authorise Dr. G. McLeod to act as assistant to Dr. Lawrie during the present session. At the same time in consequence of the circumstance that from some error or misapprehension a canvass for the chair of Surgery has been commenced and prosecuted as if the chair were actually vacant, the Senate think it right to add that in sanctioning the appointment of Dr. G. McLeod they do not mean

100

to express any opinion with regard to his professional qualifications as compared with those of the gentlemen who have announced themselves as candidates." This was as strong as could be in the circumstances. It would have been equally improper for us to have done anything directly to prejudice McLeods claims: but we felt that he was to be under great advantage in being named Lawrie's substitute.

I do hope that Lister's claims may have a fair judgment. I need scarcely say to you that I think him out of all sight the best man, & how much I should rejoice if he can be secured for our University. But testimonials are now so absurdly got up, that it must be difficult if not impossible for a person ignorant of the character of the writers to appreciate their true value. — When I lately saw Sir John Herschell's[6] and De Morgan's testimonials for Grant[7] in the Astronomy Chair I knew & said at once, that is the Man, for I knew that neither of those would write such testimonials without good reason; but among the herd who gives testimonials for Surgery chairs two thirds would require explanatory notes or testimonials to inform us who the writers themselves are, and the value of the testimonials is now almost in the inverse ratio of the strength of expression of praise in it. It is necessary therefore that men whose opinions may be relied on should assist Sir C. Lewis in his judgment and I hope there will be ways of bringing this about. I have of course like the rest of my Colleagues the deepest interest in securing the best man: and it seems very strange that parties who have no real interest in the University should be permitted to interfere upon false grounds and warp the judgment of those in authority. It is hard to say how much of political chicanerie there may be under the semblance of liberality in what is going on here. Neither of our members are very secure in their seats and I can easily imagine that they will be glad to find means to keep quiet at least a portion of their tory opponents.

I have made it a rule as much as possible to abstain from interfering in connection with appointments to our own University and I have therefore refused testimonials to all the Candidates who applied to me. I am very unwilling that my name should be used in any way connected with the Canvass: but, I think it only justice that the course of circumstances here should be rightly known and I hope you may have some opportunity of farthering the best appointment for us.

I am ever Dear Sharpey
    most sincerely yours
        Allen Thomson

[1] In November 1859 the Surgery Chair at Glasgow became vacant as a result of the death of James Adair Lawrie.

[2] Walter Buchanan (1797–1883) and Robert Dalglish (1808–90) were the two MPs for Glasgow in 1859. Dalglish, in particular, was noted for his Radical views.

[3] George Husband Baird Macleod (1828–92), Glasgow surgeon. He became Professor of Surgery at the Andersonian College in 1859, and eventually succeeded to the University Surgery Chair in 1869.

[4] Norman Macleod (1812–72), noted preacher and father of G. H. B. Macleod. He was Minister for the Barony Parish in Glasgow.

[5] George Cornewall Lewis (1806–63), Home Secretary in Palmerston's government.

[6] John Frederick Herschel (1792–1871), astronomer and natural philosopher.

[7] Robert Grant (1814–92) was appointed Professor of Practical Astronomy at Glasgow in 1859.

64

The College, Glasgow,
27th November 1859
My dear Sharpey

Though I would gladly have kept out of direct interference in the Surgical Canvass I feel it of such importance and that Lister's qualifications are so superior that I have forced myself to it and have written to Lord John[1] such a letter as I think if he is disposed to use it will do some good with Sir C. Lewis. I enclose a letter to yourself which of course you will regard as private: but you will observe it is written in such a manner that if you see it can be of any use with any influential persons to whom you have access you may privately employ it in the Cause. I have also written to Benjn Bell[2] in Edinburgh a letter the same purport, which I have some expectations he may shew to the Ld Advocate.

They are all different a little but touch upon the interference of the Glasgow members, which I hope may set up the jealousy of the men in power. I fear the effect might not be great were parliament meeting, but in the recess we may hope for something.

I must take another time to write you on other matters as I have been intending for some time. I have a larger class by 25 than last year & am very busy.

Ever sincerely yours
Allen Thomson

[1] Russell was Foreign Secretary in Palmerston's government.
[2] Benjamin Bell (d. 1883), Edinburgh surgeon.

65

Edinburgh
2nd Jany 1860
My dear Sharpey,

I find Lister's election stands just as I was supposing. There is a hitch at head quarters in consequence of the pressure of the Members for Glasgow. Lister has both the Lord Advocate and the Home Secretary in his favour, and yet the latter is so pressed upon by the Glasgow people (Members chiefly I suppose) that he refrains from making the appointment.

I learned last night that a reference has been made to the Glasgow Members. Dr. Cowan[1] was the authority and I think it very likely that he had it from Mr. Dalgleish the Member himself, so there is some truth in the report, though the terms in which the reference has been made may (as I hope they are) be exaggerated. I have got the same information this morning from a reliable source. It is very unfair to the Ld Advocate and to us also. But I hope something may yet be done to neutralise the influence. Of all things the best would be a reference to Brodie whose clearness and decision would confirm Sir C. Lewis in his selection of Lister and make him complete the appointment at once.

As the bad news kept me awake I penned this morning the arguments as they occurred to me on the accompanying papers in a letter to you, in case you may have it in your power to do any thing. I have seen Syme and we do not know of any other or better channel than through Sir James Clark and I would hope that through him or otherwise you could get a reference to Brodie made.

Of course after the scrapes I have already got in to I should like to keep out of view. I am aware too that my letter is a dreadful long rigmarole: but it contains arguments which if *properly condensed* might I think weigh with Sir George Lewis. Poor Lister who was feeling somewhat sure of success will be dreadfully disappointed and I shall not be less so if we fail.

Pray write me soon and believe me ever yours

Allen Thomson

You will understand that the subject of a difference between the Ld Advocate and the Home Secretary or the fact that the latter has not acted on the Advocate's recommendation must be kept to yourself. I have it quite confidentially —

We return to Glasgow tomorrow morning[.]

Johnny is rather better.

---

[1] John Black Cowan (1829–96), lecturer in medical jurisprudence at Anderson's College and Assistant Editor of the *Glasgow Medical Journal*.

66

The College, Glasgow[1]
2nd Jan^y, 1860
Private
My dear Sharpey,

Since I last wrote you I have learned that it is asserted in Glasgow that Sir George C. Lewis has consulted or deferred to the Members of Parl^t for the City for a decision or advice as to the fittest person to be appointed to the Chair of Surgery in the University. I still indulge a hope that this statement may not be correct or at least may be exaggerated: but as it came to me through a source from which I thought I could trace it to one of the members himself, I feel extremely anxious as to the result of the nomination to this chair. I have always heard Sir G. C. Lewis spoken of as a man of so much independence of thought and action, and of such scrupulous fairness, that I think it is possibly from the very love of justice that he has been led to adopt this plan of endeavouring to remove the doubts from his own mind, imagining that he will thus obtain a fair exposition of public sentiment upon the subject: Or it is as likely that he may have been pressed upon by the Glasgow members themselves to adopt this course by their representation that the Glasgow public have decided views upon the subject. This I believe to be altogether groundless and can only be brought forward for party purposes. We already know well how much parliamentary influence may be brought to bear on questions of this kind with which it ought to have no connection. It is most desirable for the sake of the University of Glasgow in relation to the present appointment, and in the interest of the Scottish Universities in general that a representation should be made to Sir George Lewis which might place this matter in its true light and remove from his mind the idea that the views advocated by the members are those of the community at large, or such as deserve any attention in connection with the interests of the University — Perhaps you will be able to advise us as to the manner in which such a representation could best be made to reach Sir George Lewis.

The question itself seems a very simple one. The University and the public wish to obtain the fittest and most eminent man to fill the chair, the person who will best discharge the immediate duties of a public teacher, who is best fitted by his character, conduct and manners to maintain the influence and dignity of the professorial office, and who by his ability, laborious research and scientific spirit and capacity has the greatest power to spread sound views and to extend the boundaries of Surgical science. *We* know quite well that Lister is the only man in the field who possesses *all these qualifications combined* and we know that he *possesses them in a high and unusual degree*. So certain are we indeed from our means of knowledge of his superiority that we are surprised that others are in any doubt as to the selection of the best candidate. But when we look to the testimonials (those dreadful deceits of modern times) and to the secondary influences which are habitually brought to bear upon patrons on such occasions, it must be confessed that it is not wonderful that in a matter where it is almost impossible for them to form any direct opinion for themselves, they should be involved in perplexity —

How then is that perplexity to be removed, or how may the patrons judgment be assisted? [S]urely by a reference to persons who are known to be capable of forming a judgment and to those parties who have the deepest interest in the excellence of the appointment. The Scottish University Act[2] seems to me to point distinctly to a solution of this difficulty which will be available hereafter if not at present, in the establishment of the University Court which is so constituted as to represent in an impartial manner all the University interests, and which indeed is now vested with the patronage formerly in the hands of the Senatus. *A reference to this Court would be a simple and fair thing in such a case as the present.* Perhaps it might be said that the University Court is not yet fully established, nor ready at this moment to respond to such a reference. If so, there is another course equally good, and perhaps simpler, and one which on public grounds may be considered as most eligible. I mean a reference, in the case of a Medical chair to one or more of the General Medical Council. Sir Benjamin Brodie as the first Surgeon of the day and President of the Council, and Sir James Clark as one of the Government Nominees in the Council and well acquainted with the system of medical education pursued in the Scottish Universities, are both of them men whose opinion would carry the greatest weight with the public, and who would give that opinion solely on the ground of the qualification of the candidates.

Why the Members of Parliament for the City should be consulted in preference to such men as these I am at a loss to comprehend. It could only be upon the plea that they are supposed to represent the opinions or wishes of the citizens of Glasgow and may have more reliable sources of information not accessible to other persons. A Member of Parliament is not a better judge of the fitness of professors of Surgery than other men. He is liable to be influenced by one or other political party or section of the Community, according to the manner in which he may have been returned. He is liable to influence those in power according to his own party bias. He has no special connection with the University which entitles him to be selected as adviser in relation to its affairs. A special member for the University would have had a title to be consulted in such a case: but if Parliament had thought the City members entitled to such reference, it would certainly have accorded some official position to those members in the University Court or Governing Body by the recent Act.

[T]here is no medical professor in the University (with the exception perhaps of Dr. Buchanan who early committed himself to the support of Dr. McLeod) who would not repudiate such a principle of selection with scorn and ridicule, and who would not declare that the fittest man should be chosen from [*sic*] the chair, quite irrespective of all local partialities or party influence.

I have no desire to detract from the merits of any of the Candidates in what I say: but I wish earnestly that means could be taken by patrons of a satisfactory kind to ascertain, not what are the wishes of this or that influential individual or class of the Community, but rather which of the Candidates really possesses the greatest amount of those varied qualifications which make a man an efficient teacher, a dignified professor, and give him eminence and distinction in the scientific department of his profession. I wish Sir G. C. Lewis could know and appreciate how anxious we are to have the best appointment made for the sake of the University and its Medical School. And if he could fully understand this I would not wonder at his then consulting the Senatus (that much suspected body) who have some title to offer an opinion upon the selection of a professor, rather than individuals whose position may give them influence and power, but who have not otherwise more means of forming a just opinion upon this matter than any private member of the Community of Glasgow.

I am satisfied that it can only be from a misrepresentation of the state of public opinion in Glasgow upon this matter that a reference to the Members for the City has been urged – But it is obvious that this may be a mere clamour in which a number have combined to serve a diversity of individual claims – in the same manner as the manifesto from a number of the medical profession. The citizens and the University will I am persuaded accept cordially and at once any good appointment that is made, and the profession, if the majority of its members feel otherwise at present (which I doubt with regard to the best part of them) will very soon acknowledge the wisdom and justice of yielding to merit alone from whatever place it may come. If the interests of the University be consulted without reference to extraneous circumstances, there will be no difficulty in filling up the appointment in a manner entirely satisfactory to all parties in the end.

I am ever my dear Sharpey,
sincerely yours
Allen Thomson

[1] Although written on Glasgow College notepaper this letter was composed in Edinburgh: see the previous letter.
[2] The Scottish University Act of 1858 provided for the creation of Courts with extensive powers to act as the governing bodies in all the Scottish universities.

67

The College
Glasgow, 6th Jany 1860
My dear Sharpey,

There seems to be considerable difficulty in knowing the exact nature of the pressure which has been exerted from this [*sic*] upon the Home Sec^y. I have a note from Lister this morning from which it appears that there has actually been a deputation of the Glasgow doctors to wait upon Sir G. C. Lewis — It is a great pity that he should be influenced and imposed upon by such blustering. It is quite certain that these gentlemen unite in their cry about Glasgow only to serve each one their own friend here: and none of them are agreed upon the best man – The members lend themselves to this cry from other motives. I have had a variety of accounts of their partizanship — The last I have got is that Walter Buchanan is not now for McLeod but for Fleming[1] and that Dalgleish is privately and personally for McLeod, but w^d take up Fleming's cause if he could carry thus the question of the *Glasgow* man being chosen.

The real truth is that a larger portion of the profession have no sympathy with this and that the University people have none, and it is really most lamentable that the Home Sec^y should thus under false representation & most absurd and futile intimidation yield up his judgment to such people who have nothing to do with it.

Pray let me know if you have heard anything.

Yours sincerely
Allen Thomson

I am quite persuaded that if the Home Secretary appointed according to his own judgment or the advice of a neutral man like Brodie no fault would or could be found. Some political influence must be telling upon him. I am told that Fleming has strong Manchester influence through his brother there.

You will be surprised at there being the doubt as to the sentiments of the Members here when I tell you most of what I have written in this & previous letters has come partly directly from their own statements to persons I have seen.

There must be a good deal of manoeuvering. The Glasgow people take their stand upon this "that the professor should be a *Glasgow practitioner*" for why? No one can tell —

---

[1] John G. Fleming (1809–79), Surgeon to the Glasgow Royal Infirmary since 1846.

68

The College, Glasgow
16th January 1860
My dear Sir Benjamin [Brodie]

I trust that the interest you take in the advancement of scientific instruction in medicine will lead you to excuse me for now addressing you on the appointment of a professor of Surgery in this University.

It appears that the filling up of the vacant chair has been delayed in consequence of a representation which has been addressed to the Home Secretary by a section of the Medical profession here, and which has been strongly seconded by one of the Members of Parliament for the City. This representation is to the effect that in the selection of a new professor a preference should be given to the claims of Candidates who are now practising in Glasgow.

It cannot be supposed that such a representation will by itself have the slightest weight with Sir George C. Lewis, who I am sure would at once reject the proposition in the abstract that the choice of professors for any of our national Universities should be limited to any locality or class. But it is nevertheless to be feared that when a Canvass for a chair has once begun, representations like these may in certain hands be made indirectly the means of favouring the interests of one or other of the Candidates, and there is no doubt that many who at another time would wholly repudiate such a doctrine, are either induced to support or are deterred from overtly opposing it by the fear of prejudicing the cause of individuals in whose cause they may feel a personal interest. But even as it is, I am glad to know that many of the most eminent men in our profession and the great majority of the intelligent part of the Community of Glasgow have no sympathy with the view which has been pressed upon the Home Secretary; and I feel persuaded that if a Canvass had not actually existed a very strong expression of opinion adverse to that view would have emanated from Glasgow.

I cannot help feeling extremely distrustful of the interference of Parliamentary men in such matters. In the present instance certainly, the opinion of the more enlightened inhabitants of Glasgow is not fairly represented, but on the contrary a view is upheld which tends to lower the character of the City for liberality and intelligence, and which is calculated to injure the best interests, and the reputation of the University. It may indeed be reasonably feared that Men who have no immediate connection with or interest in the University may be swayed by circumstances which have little reference to the advancement of learning but are closely connected with the relations subsisting between a Member and his political Constituency.

I feel assured that nothing would tend more to destroy the confidence which the University and the Community at large are desirous to place in the exercise of University Patronage by the Advisers of the Crown than the idea that any limitation to a particular class should be made in the choice of professors or that influences not directly connected with University interests should be allowed to operate in determining that choice. The Senatus of the University – as *a body and* individually have but one wish, which is to obtain the most eminent and the most desirable

Colleague that can be selected from any quarter. They have already expressed through the Principal in a letter addressed to the Home Secretary their repudiation of the doctrine of a local limitation of the choice of a professor for this chair: And I would venture to suggest that if a doubt should still exist as to the proper person to be selected the Senatus or the University Court have a claim to be consulted in preference to those who cannot be supposed either to possess so much knowledge of or to feel so deep and immediate an interest in the superior excellence of the professor to be chosen.

My medical colleagues and I are well aware that the expression of your opinion on this matter would have the greatest weight with Sir G. C. Lewis, and we would feel deeply grateful if you could assist us in procuring for the University, irrespective of local or partial prejudices or predilections, the most eminent and scientific Surgeon and the most desirable Colleague to fill our vacant chair.

I am My dear Sir Benjamin
very respectfully & truly yours
Allen Thomson

to Sir B. C. Brodie Bart.
Pres. R. S. &c. &c.

69

London
18th Jany. 1860
My dear Thomson

I have been dining with Sir B. Brodie nobody but myself being a guest — I had received your letter before leaving home & Brodie had got your Letter – but I am sorry to say he is quite averse to interfering in any way — He wishes to take no step unless applied to by the Government & evidently does not wish to be applied to — He thinks that none can with propriety make any remonstrance to Sir G. L. except those interested – e.g. the Senatus, or the candidates —

I cannot believe that Sir G. Lewis can act upon the result of such a proceeding as is now being enacted ostensibly with his sanction – ! It shows on the part of the Government a want of all consideration for the interest of the University, to allow this patronage to go as a sop to the Members – then the putting it to the vote of the "oi polloi['] of Glasgow registered practitioners – reckoning all votes equally good – pennies shillings & sovereigns all as equivalent units![1] It is a *Scandal* – and doubly so after the legislature had provided a proper tribunal to administer all patronage not royal – would Sir G. Lewis have treated Oxford or Cambridge after this fashion!

Send a letter to the *Times* with a copy of the circular – giving your name or some others equally good as a guarantee of truth, although any non de plume may be used in public.

I feel quite powerless in the matter.

What security has Sir G. L. that the *unsigned* crosses in Mr. Dalgleish's List represent the opinions of the persons to whom they are assigned?

It is very distressing – but after all the best counteraction is a very decided remonstrance from the Senatus – accompanied by an exposure of the real nature of the proceeding.

Brodie expressed himself very much pleased with your selection to the [General] Medical Council.

Ever yours most sincerely
W Sharpey

Dr. Allen Thomson
FRS.

---

[1] In January 1860, the Glasgow MPs sent a circular to all medical practitioners in the Glasgow area asking them to indicate the candidate they thought most suitable to fill the Surgery Chair. For the text of the circular see: *Med. Times & Gaz.*, 1860, 1: 122.

70

Wrote 22nd [This is in Thomson's hand.]
33 Woburn Place
20th Jan 1860
My dear Thomson

I have your letter of the 19th and I have seen Thursday's Scotsman – which is far too tender with Sir G. Lewis.[1]

I think he has behaved unworthily of the trust reposed in him, to yield to political pressure at all in such a case — I do not make him immediately responsible for Dalgleish's notable proceeding, but he had no business to make the interests of the University and of Medical Education in Scotland subordinate to the gratification of a political supporter. At the same time it is quite possible that the *Members* worked on Sir G. Lewis through some other member of the Cabinet – perhaps Palmerston himself, who I suppose does not care if the Glasgow Chairs are dealt with as "fish guts." The best of the joke is that Sir G. Lewis is allowed to have made the most effective Speech against the Ballot in the recent discussions of this subject – and through his means Election by ballot has been attempted in the case of an academical Professorship!

Sir James Graham, with all his inferiority in many other things, managed better in the affair of patronage.

To speak of another Graham (the Master of the Mint). He asked my opinion about a month ago about the Candidates, when of course I spoke most decidedly for Lister — Graham had been canvassed by some Glasgow candidate who was an *old personal friend* – not McLeod — My opinion had the effect of preventing Graham from moving *for* his friend – but in the circumstances I fear I could not ask him to be urgent for Lister – and unless he were to take up the cause keenly – I don't think his interference could have any effect.

I am told that Sir G. Lewis likes to share responsibility when he can – I fear if some Glasgow candidate is pressed on him, he will not ask an opinion (say from Brodie) as to the best, but whether Mr. Dalgleish's man is quite competent for the office – and select him if the answer be favourable.

I am glad to hear that Mr. Buchanan is sensible of the false step he has made in cooperating with his colleague in the matter – The M.P.s should not concern themselves about University matters unless of such as they are themselves conversant with – or on which they have been instructed to take part either by the University people or by persons who have grievances to complain of against the University.

But Post-time is come "So no more[?] at present but [. . .]"

Yours most sincerely

W. Sharpey

Dr. Allen Thomson

What side does William Thomson[2] take? if the right one, his exertion might do good —

[1] The *Scotsman* of 19 January 1860 published a leading article that described the Glasgow MPs' ballot of the local medical community as a "discreditable bit of jobbery", but which absolved the Home Secretary of blame for the manoeuvre.

[2] I.e., William Thomson (1824–1904), the future Lord Kelvin, Professor of Natural Philosophy at Glasgow.

71

College, Glasgow
16th Dec. 1860
My dear Sharpey

I am ashamed not to have written you before this: but as usual and as you know full well the early work of the session absorbs one's time and attention so much as not to favour or dispose to letter writing.

You will have heard I daresay from Syme or some other quarter of Lister's success which has been most gratifying to me. He has actually 180 in his class more than forty above the number of last year and larger than the Surgery class has been for long. His lectures give entire satisfaction to the students and they listen to him with the closest attention. Then he is intensely interested & highly pleased himself and his Colleagues are all quite taken with his pleasing manner and satisfied with his success. He had had a number of consultations & operations from different practitioners so that in practice also his prospects are good. And there seems to be no doubt that he will get the next Surgeoncy in the Infirmary —

My class is also large, in fact the largest I ever had, owing to the very large number of entrants of last year and a good continuance this year. I have 253 in all – so that I had to get some additional seats placed in my classroom. The number of beginners last year was about 110 – this year it is about 100, and I daresay it may come down next session as I do not think the present plethora is to last. At the same time I think our school will keep its place well as compared with Edinburgh in consequence of our great advantages for dissection.

I have heard nothing from Koelliker nor have I seen his Embryology,[1] altho' I see the first part advertised as published. Young Andrew Buchanan[2] is very desirous to translate the work and has I think been in communication with Koelliker about it. He is somewhat priggish and confident, but on the whole much improved by his travels, and he has recently taken off his beard, which is a sign of increasing good sense.

I am looking up the larynx and voice as well as the ear & hearing in connection with a public lecture I have to give soon in the Corporation Rooms. Can you tell me of any thing good on these subjects within reach. Is there anything on the function of the Cochlea since Cortis'[3] observations. I have been looking at a paper by Mr. Garcia in the Roy. Soc. proceedings on the singing voice.[4] Will you tell me who he is and whether you think his observations can be trusted. He is not very clear about falsetto notes and I cannot make up my mind on this subject. I think the recent Germans such as Ludwig & Ecker[5] are wrong in supposing that the posterior part of the glottis forms an open triangular chink during vocalisation: for though it is true that the basis of the arytenoid cartilages cannot be brought quite close together, this part of the glottis is quite shut by the approximation of the upper part of the cartilages — Most authors seem to me to have fallen into error chiefly by considering the actions of the muscles individually; whereas they must be taken together. It is to me quite clear that vocal position in ordinary voice is given by the crico thyroid, crico-arytenoideus lateralis and the arytenoid muscles: but I cannot make up my mind about the action of the thyro-arytenoid. Garcia's observations as to its forming arches or loops on the outside of the vocal cords are curious and as far as I can see correct, and I do not think that the action which they may have on the vocal cords has been sufficiently studied.

Our holidays will begin I think on Saturday the 22nd when we shall go to Edinburgh for some days. I wish much you could find time to write me soon as we are wearying to hear of you.

Mrs. Allen has had a curious neuralgic affection of the arm, at first very painful, but now only wearying & troublesome – she tried quinine first & is now taking Colchicum, but not with any marked effect.

I send enclosed a print done here of my negative portrait of you – from which I removed the halo, and which has been done in the vignette style to get rid of the scratches at the side. I think it the best we have got of you & beg Miss Colvills acceptance of it with my best regards.

Ever most sincerely yours
Allen Thomson

[1] Albrecht von Kölliker, *Entwickelungsgeschichte des Menschen und der höhern Thiere*, Leipzig, W. Engelmann, 1861.
[2] Andrew Buchanan (1834–65), Glasgow medical practitioner.
[3] Presumably F. Cortese, 'Delle recenti scoperte degli anatomici sulla struttura e sugli uffici della coclea', *Annali univl. Med.* [Milan], 1854, **149**: 309.
[4] Manuel Garcia, 'Observations on the human voice', *Proc. R. Soc. Lond.*, 1854–5, **7**: 399–410.
[5] Presumably Carl Friedrich Wilhelm Ludwig (1816–95), and Alexander Ecker (1816–87).

72

33 Woburn Place, London WC
4th March 1863.
My dear Thomson

I am amused to think that I should be in a position to receive excuses from any body for remissness in correspondence! No one is more conscious of standing in need of indulgence. Never the less the lapse of time since I had your last has been truly owing to the fact that the letter referred to business on which I was rather indisposed to enter – I mean that "weary Quain's Anat^y"[1] — I have talked with Walton, with Marshall & (today) with Ellis. Walton seems convinced that it would not pay him to publish a new edition on the existing plan – that it will need to be reduced in extent & illustrated with larger figures. I doubt if the idea of a separate Elementary Atlas of Anatomical Figures will be persisted in. It seems preferable to enlarge the size of the page of the Book to Royal Octavo and introduce large figures into it. Now I find that Ellis thinks Walton ought to keep the Book as it is – with more illustrations – but *not abridged* – At any rate he cannot undertake to abridge his part for then he says he would need to re-write it, and this he could not do. I am quite willing to curtail and excise or compress so far as my part is concerned – and I begin to think the book must be either reduced or extended – extended, viz., if it is to take the place of a full system. But a full Treatise will perhaps some day or other be undertaken by an association of writers (a la Soemmerring's Anat^y[2]) and therefore I am reconciled to the reduction. Still in the Descriptive Anatomy I think no actual insertion or process or foramen or branch of an artery should be left out. The details of description may be shortened but it must be understood that everything's noticed.

In my part I should omit all discussions – shorten descriptions and deal with questions more dogmatically but I see that many figures in addition to or substitution for those existing will be needed.

Now Ellis having declined, we must think of another Editor — Walton fears that you would find it irksome labour & would not come up to time. I grudge recommending anything to you that you might feel as a drag on better work – but if you could like it and think you could work on it pleasantly I should at once get him to arrange with you, to cut down and dress the text & *direct* the preparation of the *hardest* cuts. Failing you I have suggested that probably Dr Cleland[3] would undertake the duty, and it would be a recommendation that he could advise with you when required.

My view of the present state of the case is therefore as follows.
1. Walton will not keep up the book at its present bulk.
2. Ellis will not reduce his part, but willingly abandons the text to another.
3. The Figures will be needed in the book to give it a fair chance of sale, with competing publications.
4. You or Cleland – or you and Cleland –, if you were willing, might very well edit the Descriptive Anatomy – i.e. including the organs of the Senses, and also the Brain, which last was my work. Pray think of this proposal – & if you approve of enlisting Dr Cleland be so good as open the matter to him.

Yesterday & today have been days of warm sunshine – too warm I fear for the salutary discipline of the growing crops. Naturally we wish Saturday may be as good after all people are excited less I think by curiosity to see the procession, than by desire to testify interest in the advent of the young Princess – & to show good will towards her – and the Queen's family.[4]

Mary received the portraits this morning – she is not at this moment in the house but I may take upon me to thank you in her name for your attention. A Letter from Helen last week informed me of Ninian's proceedings & his sisters arrival in India – also that your windows at Moreland were in.

With kind regards to Mrs T. & Johnny —

Yours ever sincerely

W Sharpey

Dr Allen Thomson
&c

P.S.   What a mess Owen has got into! observe how he tries to *hedge*.[5]

---

[1] I.e., the seventh edition of Quain's *Elements of anatomy*, published in 1867 by J. Walton with Sharpey, Thomson, and John Cleland as joint editors.

[2] Samuel Thomas Soemmerring, *Vom Baue des menschlichen Körpers. Neue umgearbeitete und vervollständigte Original-Ausgabe*, Leipzig, L. Voss, 1839–45.

[3] John Cleland (1835–1925), Thomson's assistant in the Glasgow anatomy class.

[4] On 5 March 1863, the Prince of Wales was married to Princess Alexandra at St George's Chapel, Windsor.

[5] A reference to a controversy between Richard Owen and T. H. Huxley over whether the human brain was structurally distinct from that of apes. Although taking Huxley's side in this dispute, Sharpey had in 1862 rejected a suggestion from Huxley that Owen be excluded from membership of the Council of the Royal Society because this would appear to endorse his claims: Sharpey to Huxley, 13 November 1862, University College London MSS, Add. 227/4. 1862 107–129 (123).

73

College Glasgow
10th March 1863.

My dear Sharpey

In reply to your inquiry in yesterday's letter I may say that no time is yet known for the election of a Professor of Latin in Ramsay's[1] place for this reason – in the first place that he has not yet given in his resignation. His intention is I understand, following the course now prescribed by the Commiss[rs] Ordinance, to send in his resignation to the first meeting of the Univ[y] Court which will not be till after Lord Palmerston has been here on the 30th of this month. It must then go before the Privy Council to receive their approval before liberty to resign can be obtained; and after all that the applications of Candidates must be received and other meetings of the Court held for the election. I scarcely think it can be made before the middle or end of May.

With respect to the new edition of Quain and the Anatomical plates I had, on trying the arrangement of the thing by myself, come very much to the same way of thinking with Mr. Walton and you. I found in the first place that we could not exactly copy with advantage any existing plates, at least that we could only copy some and would need

113

either to supply or to modify others; and I found upon trial that I would not be satisfied with the most of them if I had not drawn them myself, or at least had given as much time to the preparation as would enable me to do them myself. I therefore abandon the plan to have a small system of plates suitable for students, though I think it a great pity that this cannot be had. The nearest thing to a suitable work of this kind I have seen is a new edition of Bock's Hand Atlas.[2]

As to the Textbook I am still of the decided opinion that we have not yet got the requisite students book, and that Quain contains all that is necessary if modernised properly and shortened by leaving out all words and statements that are superfluous. I would not on this account leave out any particulars in description; but I think it could be greatly shortened, and much assistance could be given by more tabulation and employing different kinds of types. The descriptive Anatomy alone should not exceed one volume. As to the illustrations I have come to think that much might be done by adopting a large size, as you suggest; And what helped to bring me to this view was the sight of a book lately published by d'Alton on the organs of motion[3] by which he means the Bones Joints and Muscles, beautifully illustrated, and the woodcuts of a size which quite satisfies me viz. a little larger than Grays. You should take a look at it. The woodcuts of the Bones are the finest things of the kind I have seen: but unnecessarily large. The muscles are rather too tame in the execution but the plan is good.

Now as to editorship, Mr. Walton and you seem to have a correct notion of me when you fear I would not be up to time. I had no doubt that Cleland would do it well and like it, and therefore I spoke to him at once; and he expressed himself much pleased with the proposal and ready to undertake it. At the same time though I cannot undertake to go through the drudgery of the whole I should like much to be associated in the work, for I feel that I have got experience of a number of things which I could made very useful in a students book, and which if I am allowed to introduce, may tend to satisfy me so much as to promote greatly the sale of the work in Scotland.

I should like therefore to have my name associated with yours as one of the editors, and Cleland should be assumed to do the hard work. He will have great advantages in doing it here with a dissecting room and good library at his disposal. I would write more on this subject. but we are just preparing to go out to see the illuminations, and I have my hands full of other matters.

It has been most gratifying to see the accounts of the procession &c. in London going off so well: And altogether the popular enthusiasm and cordial good feeling to the Queen and her family have been most delightful.[4] I have just been to a morning entertainment given by the Town authorities at which great satisfaction has been expressed and there are to be public dinners & balls of various kinds in the evening.

We are having also a torch light procession of the students to the number of about 400 which causes me some anxiety: but I hope it will all go off well.

Owen's long sentences & grandiloquent & obscure phrases are very funny: but I wonder the Londoners are not tired of the Gorilla's brain. I have found my drawings of the Chimpanzee's again: but I should be ashamed to publish now as I once intended —

With kind regards to Miss Colvill & to your self from Mrs. Allen I am

Ever sincerely yours

Allen Thomson

114

[1] William Ramsay (1806–65), Professor of Humanities at Glasgow from 1831 to 1863.

[2] Presumably, August Carl Bock, *Chirurgische-anatomische Tafeln; oder Abbildung der Theile des menschlichen Koerpers in Bezug auf chirurgischen Krankheiten und Operations*, Leipzig, Voss, 1833.

[3] Edward d'Alton, *Die Anatomie der Bewegungswerkzeuge; oder, Knochen-, Bänder- und Muskel-Lehre des Menschen*, Leipzig, H. Hunger, 1862.

[4] See note 72.4 above.

74

Wrote 4th Jan, 1864   33 Woburn Place WC.
                                                       31. Dec. 1863

My dear Thomson

In determining a time for the Genl Council meeting we did not forget the case of Professors & Lecturers – but the Christmas Holidays would have been complained of because of too short warning & perhaps interference with engagements settled beforehand for that time of year. A Monday was objectionable because Scotch Sabbatarians would be brought away on the previous Saturday. Saturday itself was liable to the risk of detention till Monday for the same reason – Supposing too the possibility of an abridgement of the Meeting till the morrow, Friday & Saturday were both objectionable. It is always uncertain how any proposed arrangement of days will affect twenty people distributed in England, Scotland & Ireland. But, after all, the time was open to suggested change.

Your Pupils cannot suffer from want of four or five Lectures – especially now as the maxim seems likely to be "the fewer lectures the better" – always excepting those on Clinical Surgery delivered in Edinburgh by an old and valued friend of mine whose name begins with J. S.

Pray let me know whether you are coming up.

You will be sorry to hear that Dr. Watson[1] definitively declines being put in nomination for the Presidency. I now see no course to take but to elect Burrows[2] or take some non Professional personage — If we could get a retired L[d.] Chancellor or Literary Judge – or some Nobleman known to take an interest in Literary or Scientific matters, I think we might do worse than choose him. Men in the Government I fear could not afford the time. Try and think of somebody of this class. On casting about in my own mind I confess I can find no one *just exactly* fitting. Then a Council of Doctors is so professional! A retired high Functionary like Lord Canning[3] or Lord Elgin[4] could such be found would suit well.

I have a note from Burrows today inclosing Watson's Letter to him with his final negative answer — Burrows thinks we must turn to some non-professional – but perhaps he does not know that several members would be well pleased to elect himself.

Lister called on Monday – we had a long chat but he is always full of his own matters – inflammation &c – so that I could get little out of him about things in general.

I have a letter from Acland[5] after he had read Syme's address[6] — The burden of it is consolidation of Examining Boards – which doubtless is most desirable – that is to say if friend Syme will allow any examinations except class examinations subject to

Andrew Wood's[7] sensible Inspectors. In London the Colleges of Surgeons & Physicians ought to form a joint Board of Examiners to confer the common Licence. There ought to be a single board for conferring *Medical* Degrees in England, the like for Scotland & for Ireland. But Oxford & Cambridge would scarcely relish this – The Scotish [*sic*] Universities still less – the great desire of their Professors is to keep hold of the Examinations. Could it only be practically brought about that regular attendance and steady attention throughout a course were made to weigh at an Examination for Licence or Degree I should be very glad to see such consummation – but what are you to do with particular teachers! must you treat all equally?

The cry of too much lecturing is in great part not well founded – it is but a cry. When I was a student the Edinburgh Courses were just as long as now. Hope[8] always cut into May. Barclay[9] towards the end joined his morning & evening course so as to make two progressive lectures daily and at last had a third one in the morning – & goodness knows he wasted much time in talking about muscles &c. His account of the Ear had to be comprized in half an hour so wasteful was he of time in the earlier parts of the course. Syme talks of "refinements" & "minutiae" and the like – but in this way an old man might always stop progress – like Dr. French[10] of Aberdeen who complained of a Young Man called Davy who was making troublesome innovations in chemistry. I cannot suppose Playfair's[11] Course is longer than Hope's and therefore I infer he makes a *selection* from the multifarious matter of modern Chemistry. I am quite agreed that certain Scientific branches auxilliary to Medicine are not to be taught or at least examined in the same detail as fundamental subjects such as Anatomy — But to tell the truth I would rather leave out certain things & allow the Student to learn the rest well, than try to squeeze all into smaller compass. Botany has no business as a branch of Medical Education. It came in *originally* into the *M.D. curriculum* because it comprehended Materia Medica I certainly should not exact both Botany & Zoology — A reasonable foundation in at least one classificatory science I think is proper – but it is unreasonable to exact details of particular parts. — Then Materia Medica & Midwifery should be Summer Classes. Lastly – there should be no double attendances *required*, on any lectures *whatever* – not even on Anatomy. Some students will find it to their advantage to attend twice or even more – others will fix or extend their knowledge in other ways but by not exacting repeated attendance there would be ample time left for practical work – especially if the winter session were relieved as in London by the transference of certain classes to the Summer.

There is however a great fallacy running through all this outcry namely – that whatever time is saved from lectures is diligently employed in private study! or in Clinical instruction. To many students the Hospital is a mere lounge. Clerks[,] Dressers & House Surgeons profit by it – the bulk of students learn very little.

I am glad Cleland has succeeded – I had little expectation of Waller's Candidature,[12] but it was suggested by a Dublin man – and I wish much he could get *something* – You rightly judge that Anatomy is not his forte, but he would have got on. However I don't think the election lay between him & Cleland & trusting you may find an adequate Successor & laud his success. I have barely room to wish you a happy new year

11 p.m.   W. Sharpey

[1] Presumably Thomas Watson (1792–1882), Consultant Physician to King's College Hospital and a former member of the General Medical Council.

[2] George Burrows (1801–87), Physician to St Bartholomew's Hospital and Treasurer of the GMC.

[3] Presumably Charles John Canning (1812–62), who retired as Governor-General of India in March 1862. However, Canning had died in London in June of the same year—a fact which seems to have escaped Sharpey.

[4] James Bruce, eighth Earl of Elgin (1811–63), a former Governor of Jamaica and Governor-General of Canada. He was appointed Governor-General of India in Canning's place.

[5] At this time Acland was Regius Professor of Medicine at Oxford.

[6] See 'Mr Syme on medical education', *Br. med. J.*, 1863, **ii**: 692–4.

[7] I.e., Andrew Wood, the Inspector of Anatomy in Scotland.

[8] Thomas Charles Hope (1766–1844), Professor of Chemistry at Edinburgh from 1795 to 1844.

[9] John Barclay (1758–1826), a private lecturer in anatomy in Edinburgh.

[10] George French (1765–1833), Professor of Chemistry at Marischal College, Aberdeen from 1793 to 1833.

[11] Lyon Playfair (1818–98), Professor of Chemistry at Edinburgh from 1858 to 1869.

[12] Cleland left his post as demonstrator to Thomson's class in 1863 to become Professor of Anatomy at Queen's College, Galway. Evidently, Augustus Volney Waller (1815–70), sometime Professor of Physiology at Queen's College, Birmingham, was also a candidate for the position.

75

33 Woburn Place 12th July 1864

My dear Thomson

I had a Letter from Koelliker on Saturday to which I replied by return of post. He announced his intention of being here on his way to Scotland soon after 8th of August.

I told him we expected a visit from you & Mrs. T. about that time, but that never the less we should be happy to receive him & his party as you would find quarters with the Grahams or some other friends.

Now I find you have abandoned your plan. I declined an invitation from Dr. Paget[1] to go to Cambridge. The occasion of a "Gathering" is the very worst for seeing such a place as an University Town – for the "natural life" of it is gone for the time; and a collection of General Practitioners with a sprinkling of M.D.s forcing you to attend morning meetings & evening converzationes – is bad employment for vacation time. I suppose the Koellikers will stay with us two or three days at farthest. When you write to Würzburg entreat of them not to delay coming – for otherwise they will probably find the town empty of their friends I know that Busk is going to Gibraltar to explore Caves; but *when* I do not exactly know. Mary has written to Mrs. Carpenter to learn about their movements – (also to Mrs. Busk) & she will write as soon as she gets an answer.

I propose that Mary should go down to Skelmorlie with the Koellikers. I am sorry to say that *as usual* I have yielded up my time to others so much, or squandered it away – that I *must* stay in London to do something towards the Histological part of *Quain's* Anat^y. How I hate the title! I have decided to stay here till the middle of September – and make a very short visit (at least compared with my wishes) to Scotland this year.

So Lister will in all probability leave Glasgow, and you will be again in the hands of the Home Office.[2] I hope you will find a Successor who will be marked for superiority in Something – If not an original contributor to the Science of Surgery – that he should be a Good Operator – or Good Lecturer – and especially a Good man.

Yours always most sincerely

W. Sharpey

P.S.   Mary had a letter from Mrs. Mylne the other day dated from somewhere near Naples. They are about to return to England – and mean to take lodgings for a short while in London near the west end parks.

Will Koelliker go to the Bath Meeting of the Brit Assoc[n] – which begins on the 14th of Sep[t]? Lyell[3] to be President.

---

[1] George Edward Paget (1809–92), Linacre Lecturer on medicine at St John's College, Cambridge.
[2] Lister did not, in fact, leave his Chair in Glasgow until 1869.
[3] Charles Lyell (1797–1875), Scottish geologist.

76

33 Woburn Place London WC
25 Oct. 1864

My dear Thomson

It has seemed to me worth while to draw your attention to a muscle of the hand said to be often present (by Henle and by Mr. Wood Proc. Roy[l.] Soc. Vol XIII p 302.)[1] It is a *palmar* interosseous muscle going to the dorsal aponeurosis of the thumb.

I am much pleased with the figures of the vertebral muscles. The old ones looked very pretty – but the new ones with the muscles raised to show the steps [?] of attachment are much more useful. Indeed they seem to me unique.

I think that in the "Advertisement" prefixed to the part of the work which will first appear – it may be expedient to give some explanation in reference to the figures. — It might be stated, for example, that the question of marking the objects with the names (putting the names of the parts on the figures) had been well considered – but as this can seldom be effected with exactness, the spot being often much smaller than the title, and as it tends to obscure and disfigure the delineation – it was not adopted — Then – that great care had been bestowed in placing the reference letters & numbers so as to make the explanation as descriptive as possible. Also that the figures of the bones are taken from the actual object & the figures of the muscles in many cases finished from actual dissections. &c. &c.

Our new entries at the College this year are much the same as last – that is *small* – and this being the third year of small entries it tells unfavourably on the whole numbers. I know nothing of the other schools save from the statement of general deficiency in the Lancet.

What a dance A. Wood has led us into on account of the alleged refusal of the Irish Poor Law Commissioners to employ Surgeons who have not a special *licence* in Midwifery. It turns out that they accept a *certificate* from any Licensing Body that *examines* in Midwifery – & that last year two such certificates were actually accepted as sufficient from the Coll. of Phys. of *Edinburgh* signed by Dr. Burt![2] In the whole of that business the Scotch have been thoroughly "licked" by the Paddies.[3]

Yours always sincerely
W. Sharpey

Dr. A. Thomson

[1] John Wood, 'On some variations in human myology', *Proc. R. Soc. Lond.*, 1864, **13**: 299–303.
[2] Presumably John Graham Macdonald Burt, a Fellow of the Royal College of Physicians of Edinburgh.
[3] These remarks appear to refer to some discussion in the GMC, but I have been unable to discover their context.

77

33 Woburn Place WC
London Feb 18 1865

My dear Thomson

As regards Carter[1] (of Leamington)'s researches – of course you know they are intended to *prove* by injections a doctrine somewhat the same as has been entertained by Virchow, Heidenhain, Recklingshausen[2] &c. – A communication between blood capillaries and lymphatics through the medium of so called plasmatic channels – at least this is nearly the pith of the matter. I have seen Carter's drawings, and Marshall & Beale[3] examined carefully a selection of his preparations sent up here for the purpose. At that time I was otherwise engaged and did not investigate them. M. & B. thought the evidence afforded by the preparations not satisfactory.

You and I know well enough that the mere passage of injections from one class of vessels to another is no proof of natural intercommunication. These *transparent* injections, such as Dr. Carter used are so liable to transude and stain extravascular objects, that, although Dr. C. thinks this was sufficiently guarded against in his experiments, I have very little faith in such proofs. We could in fact get appearances from mere immersion in carmine & other coloured solutions, showing channels filled with the material – i.e. their contents coloured by it. A nucleus in the centre of a so called plasm cell or connect[v.] tissue corpuscle will become stained while the surrounding matter is comparatively untouched. But I am satisfied the doctrine that the connect[v.] tissue corpuscles are really ramified *cells* must be given up – if it is thereby meant that they have a cell wall and that their branches are tubular. I believe Virchow himself is now driven to admit that the bodies in question lie in interstices of tissue which give the appearance of a branch cell with cell wall and contents. The lymphatics may begin in these interstices, and the nature and *motion* of their fluid may be influenced by the plasm *corpuscles* which lie in these interstices. There appears to me *much less* evidence in favour of a communication of these intersticial passages with the blood capillaries. In any case I don't think Dr. Carter has added anything to the existing evidence for either doctrine.

119

As to Dr. Beale's "paths of the nerve currents"[4] I will not venture to speak till I have seen more of his preparations.

He tortures the things so much with chromic acid, glycerine &c. – that one feels uncertain as to the appearances of fibres. You remember the mistake of Brücke (or was it Hannover?)[5] who after steeping an eye in a chemical solution (alum, or sugar of lead?) found the vitrous humour to be concentrically laminated like an onion. Now suppose a nerve fibre thus acted on – in section [drawing here, see figure 7] – in profile thus, [drawing here].

I by no means would have you to understand that I have this or similar doubts about all Beale's preparations – but in some others of them the nerve fibres are shown only with very high powers, and they seem to run out at the end into fine threads which look like clear *transparent* lines or streaks; and indeed unless I make myself familiar with "twenty fifths" and perhaps with "fiftieths" – I could scarcely pass a judgment. Moreover you know better than I how necessary it is to work out the observation for yourself from the beginning, & how unsatisfactory it is to be shown perhaps a single fortunate specimen to prove a difficult point.

I think you may take it for settled that the Med Council will meet on the 4th of April at which time (or sooner if you like) we shall be prepared as usual to receive you; and Mrs. Thomson if she will come. I have written to Giraldes to say that we all assent to his proposed translation.[6]

Yours always sincerely
W. Sharpey

Dr. Allen Thomson

[1] Thomas Albert Carter, 'On the distal communication of the blood vessels with the lymphatics; and on a diaplasmic system of vessels', *Proc. R. Soc. Lond.*, 1864, **13**: 327–9.

[2] Rudolf Carl Virchow (1821–1902), German pathologist; Friedrich Daniel von Recklinghausen (1833–1910), German histologist and pathologist; and Rudolph Peter Heidenhain (1834–97), German physiologist.

[3] Lionel Smith Beale (1828–1906), Professor of Physiology and General and Morbid Anatomy at King's College, London.

[4] Lionel Smith Beale, 'Indications of the paths taken by the nerve-currents as they traverse the caudate nerve-cells of the spinal cord and encephalon', *Proc. R. Soc. Lond.*, 1864, **13**: 386–92.

[5] Ernst Wilhelm von Brücke (1819–92), German physiologist; Adolph Hannover (1814–94), German histologist.

[6] Joachim Albin Cardoza Giraldès (1808–75), Spanish medical author. These remarks seem to refer to a proposed translation into Spanish of Quain's *Anatomy* by Giraldès; if so, the project appears not to have been executed.

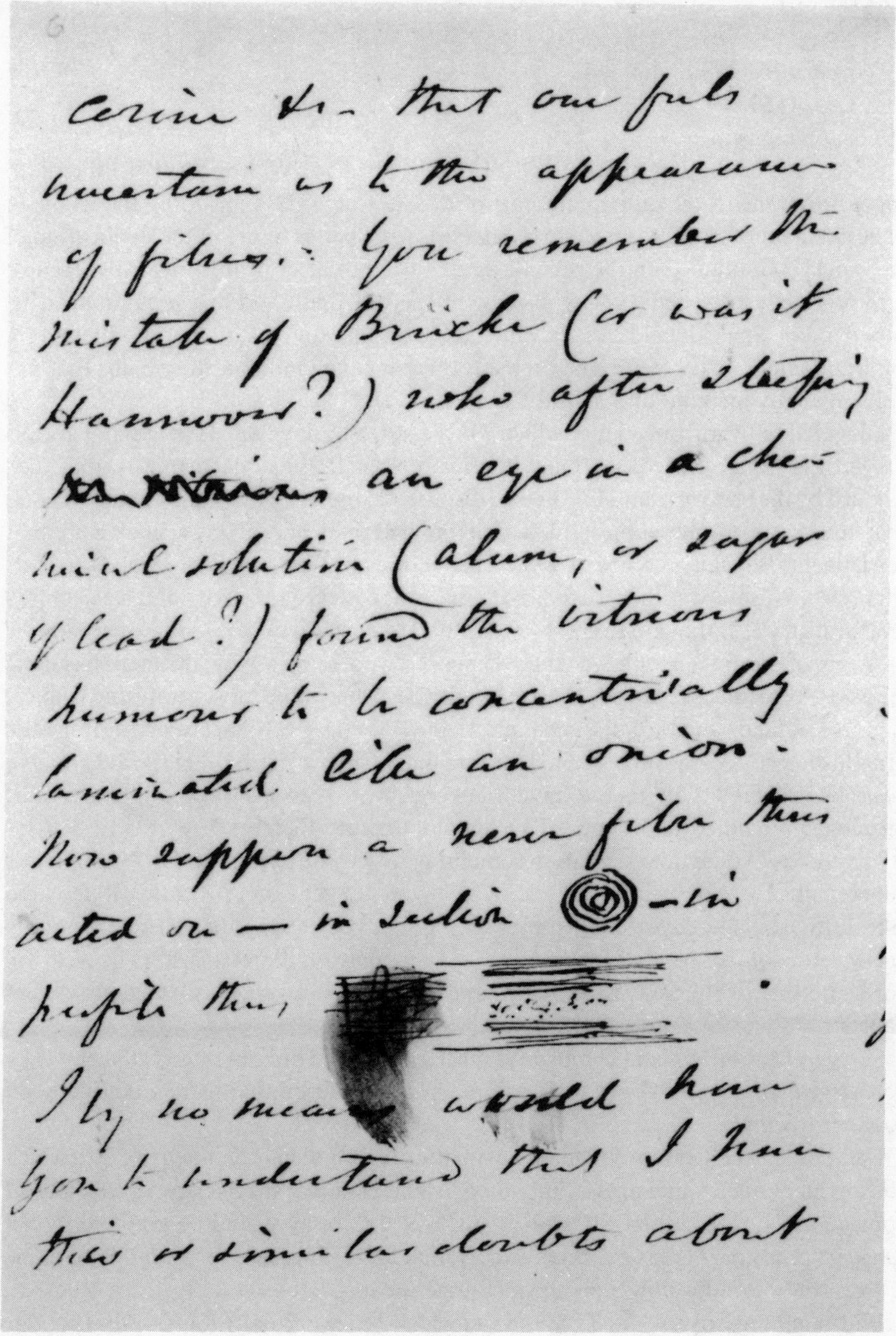

certain &c. that one feels
uncertain as to the appearance
of films. You remember the
mistake of Brücke (or was it
Hannover?) who after steeping
~~an vitreous~~ an eye in a chemical solution (alum, or sugar
of lead?) found the vitreous
humour to be concentrically
laminated like an onion.
Now suppose a nerve film thus
acted on — in section ⊚ —in
people thus,

I by no means would have
you to understand that I have
these or similar doubts about

Figure 7.

121

78

33 Woburn Place, London WC
10th Dec. 1865
My dear Thomson

Since I last communicated with you I have sent to Mr Walton all that I intend to appear from me in the outcoming Part of the Anatomy. It is including the whole of Bone and a bit of Muscle. I was sorely at a loss for a better figure of the ossification of the Embryo cranium than the present figure 36' – but after trying one or two people here was obliged to send in my copy with the old figure, adding a caution on its imperfection. Since then I have got Marshall to try his hand at a preparation I made & the wood cutter, after seeing the preparation, has taken in hand Marshall's figure. I sadly missed your kind and skilful hand.

I am glad to hear from a letter which Mary had yesterday from Helen, that Cowan[1] has got the Materia Medica Chair. It is just one of those subjects that is not much pursued by the best men; and I should be disposed to prefer a *good* man who had a good deal to get up in the subject to a dull or mediocre one already known in that department. We had a choice of several some three or four years ago, when Garrod[2] left us; Murchison[3] was then a candidate, and as he is clever and active and had lectured on Chemistry & Botany in India, I felt inclined to push his claims – but eventually we took one of our own people – Sydney Ringer[4] – who being a very able man has done the Materia Medica work very well, & besides he is one of the most promising Clinical men in England, and withal a very nice fellow; the last can hardly be said of little Murchison who has the reputation of quarrelling with everybody. (This is said *privatim loquens*). I feared you might have Fleming[5] who was at Cork and latterly at Birmingham palmed upon you – The Aconite tincture Fleming.

I really wish you would send us in something by way of Report on Dr Wilson Fox's Paper on the Development of Muscle.[6] Could you not get Lister (the other Referee) to take the first trouble & you could join in his report if you approved it. There will be only one Meeting of Council before the New Year, namely on the 21st Inst. (The next will not be till the 3^d Thursday of January) and we are anxious to get through arrears of Papers.

Sorry to hear of Ninian [Thomson]'s mishap – but the broken bone will mend. The non survival of their first born will no doubt be a great disappointment – but there will be more to come.

I suppose Frauenfeldt of Vienna sent you the account of his "Samuelreise"[7] in which you are mentioned – and also a diploma of Membership of his Society, in which too I hope he had not forgotten Mr Robertson[8] who I daresay would be gratified by the *honour*. For my part I may say unaffectedly, that these things are very indifferent to me & I can only wonder how some people prize them.

With kind regards to Mrs T. & John in which Mary cordially joins – Always your sincere

W. Sharpey

[1] John Black Cowan, who became Professor of Materia Medica at Glasgow University in 1865.

[2] Alfred Baring Garrod (1819–1907) Professor of Materia Medica, Therapeutics, and Clinical Medicine at University College London until 1864.

[3] Charles Murchison (1830–79), former Professor of Chemistry at the Medical College, Calcutta (1853–4), was then lecturer on botany and curator of the museum at the Westminster General Dispensary. From 1860 he was Assistant Physician and pathologist at the Middlesex Hospital.

[4] Sydney Ringer (1835–1910), Professor of Materia Medica, Pharmacology, and Therapeutics at University College London (1862–78).

[5] Alexander Fleming (1824–75), Professor of Materia Medica at Cork and from 1858 at Queen's College, Birmingham.

[6] Presumably Wilson Fox, 'On the development of striated muscular fibre', *Proc. R. Soc. Lond.*, 1865, **14**: 374–6.

[7] Presumably Georg von Frauenfeld (1807–73), Austrian naturalist and traveller.

[8] Possibly Douglas Argyll Robertson (1837–1909), an extramural lecturer on diseases of the eye in Edinburgh.

79

Huntingdon
Dec. 24th 1866
My dear Sir,

I had the honor [*sic*] of making your acquaintance several years ago at Aberdeen at the British Association. May I venture to presume upon that to ask your assistance in the following matter.

Through the resignation of Dr. G. Harley the post of Instructor in Practical Physiology and Histology at University College London has become vacant, and I am about to become a candidate.

For the last few years I have been busily engaged in practice in the country, and have had few opportunities for original research, but have never been able to keep myself from Physiology. For domestic reasons I have been obliged to remove from Huntingdon and give up my practice, and the post at University College having offered itself, I have determined to apply for it, should I succeed, my intention is to devote myself almost entirely to Physiology. If you can honestly say a good word in my favour I shall esteem it a great favour.

, I may add that Dr. Sharpey views my candidature very favorably [*sic*].

Believe me dear Sir
Yours faithfully
M. Foster[1]
Prof. Allen Thomson.

[1] Michael Foster (1836–1907), future Professor of Physiology at University College London and later at Cambridge. For particulars of his career up to 1866 see the next letter.

80

Foster's Address at present is 3, Lansdowne Villas, Bournemouth

---

Wrote 4th Jan [This is in Thomson's hand.]
33 Woburn Place, London
2d. Jan 1867
My dear Thomson

I have just recd. the Draft of Michael Foster's application – The chief points are. B.A. of the Univ. of London gained the Classical Scholarship & gold medal —

At Univ. Coll. gained Gold Medals in *Chemistry* and *Physiology* & 2d place in Comparative Anatomy.

Was Assistant Curator of our Museum till obliged to go to the East for his health.

Whilst in practice in Huntingdon has carried on physiological investigations and published Papers.

List of Papers

On the beat of the Snail's heart. Brit. Assoc. at Aberdeen 1859.

On the effect of freezing on the physiological properties of muscles. Proceedings R. Socy. 1860

Contribution to the theory of Cardiac Inhibition Brit. Ass. at Oxford, 1860

The Coagulation of the Blood. Nat. Hist*y* Review April 1864. A Critical Essay, with corroborative original Experiments

On the existence of glycogen in the Entozoa Proc. R. Soc*y* 1865

On Amylolytic Ferments Journ. of Anat. & Physiol. No. 1 Nov 1866

Articles on *Nutrition*, *Respiration* and the physiological subjects in Watt's Dictionary of Chemistry

The life & writings of Sir B. Brodie North Brit. Review Sept. 1865

The Elements of Muscular Strength Fortnightly Review Sept. 1. 1866

And other Papers.

You would do him a great favour if you would send him a Testimonial with your earliest convenience.[1]

In the course of last night Snow has fallen a good deal more than Ankle deep. —

I have just been shocked – very deeply – by hearing that Dr John Russell[2] has killed himself —

Wishing you & yours many happy returns of the New year
            Yours always sincerely
                W. Sharpey

[1] Foster's application for this post was successful; it marked the start of his career in physiology.
[2] John Rutherford Russell, an Edinburgh MD resident in London, was reported to have hanged himself in January 1867. See: *Br. med. J.*, 1867, **i**: 15.

81

Hampstead, London N.W.
15th November 1868
My dear Thomson,

Conscious of blameworthiness I was ashamed to be reminded in a note from John Marshall last night that I had so long delayed communicating with you. We were none the less interested, however, in your great doings in Glasgow, and eagerly read the accounts of your grand Ceremony.[1] The fault was that one who like yourself, has from the first to last worked so steadily and effectively in promoting the undertaking should not be singled out for special honour. Not that I wish to see you be-k-nighted like the worthy Provost – but your Portrait or Bust ought to occupy a conspicuous place in the great Palace; and I trust it will do so in time.

I was more than two months on the Continent – in Switzerland, going as far as the Engadin. I profited considerably by my stay in the higher regions – but I made a wrong move at last in going to [. . .] which was hot and relaxing, & which there & in Geneva I was never far from a severe Cold. But in Geneva I was much pleased to meet Lombard[2] who invited us to his house & entertained us hospitably at dinner. We also visited Claparède[3] who in Summer lives a couple of miles out of Town – He is presumably known to you I think.

From Geneva we went for three days to Aix les Bains in Savoy & came down to Lyons on the Rhone Steamer. Ollier[4] was found very kind – He is Surgeon to the great Hotel Dieu of Lyons – but I was wearying to get home so we made but a very short stay & came home by Paris. I feel very well just now. I take well with dry *frosty* weather, of which we have had a fair share.

Our classes both Medical & Arts are good this year. I have 68 new entries (which means mostly perpetuals) last year at the end of the Session I counted 56. I presume Glasgow & Edinburgh must also be good this year, but I have heard nothing. I shall not let so long a time pass before writing again – Mrs Storrar is now wonderfully well.

With most kind regards from Mary & myself to you & yours – I remain My dear Thomson

Yours most sincerely
W Sharpey

---

[1] A reference to Glasgow University's transfer to new premises at Gilmorehill. Thomson played a major part in the planning and execution of this move.

[2] Presumably Henri Clermond Lombard (1803–95), Swiss physician and zoologist.

[3] Presumably Jean-Louis-René Claparède (1832–71), Swiss embryologist and zoologist.

[4] Louis-Xavier-Edouard-Léopold Ollier (1830–1900), Chief Surgeon at the Hôtel-Dieu of Lyons from 1860.

82

Wrote 10th April. A.T.

Hampstead, London N.W.
29th March 1869

My dear Thomson

As on many former occasions I have to begin with an acknowledgment of my defaults in correspondence, and an expression of gratification for your last Letter.

We shall be delighted to see you here. You say about the 24 April (the 2d Soiree at the RS). I hope you will not make a hurried visit & that Mrs Allen will be with you. You will I hope see my Nephew Wm Henry[1] whom we expect to arrive from Bagdad [*sic*] by the end of April. He has now been thirteen years abroad. He held the place of Civil Surgeon at Bagdad in lieu of a Dr Wood who was at home on leave & did not mean to return – but Mr W. has only lately intimated his retirement, and my Nephew is now confirmed in the appointment – so that it will be open for him when he returns to the East.

A meeting of the English Branch Council is summoned for the 31st (Wednesday next) and the time for assembling the Genl Council will not improbably be discussed – In which case I will let you know what is wished for in this end of the Island.

Storrar[2] is dead against Syme as suggested for the Presidency [of the General Medical Council], his objections go beyond non-residence, which *seems to be* the chief or only objection entertained by others. Had we had an unexceptionable choice in London – the same objection to Syme would have prevailed with me. Last year I thought of Caesar Hawkins,[3] but understood (from the President I believe) that he was unwilling to be brought forward. It now appears that this was a misapprehension; but in the mean time I declared my intention of supporting Syme and do not regret I did so.

The chief objection to non-residence is the short notice at which the President is generally required to hold interviews with the Members of the government. But as the request for such interviews almost always proceeds from the President himself there would be opportunity for arranging them to suit his time. After all it could only make a day's difference – and then the difficulty will not arise when the Council is in Session. I met Playfair the other day who strongly supported the election of Syme as a source of fresh interest and activity to him which would be most salutary after his recent bereavement.

We have had (as you know) a very mild winter – but of late we have been visited with cold frosty or wet winds – very trying. For myself I have kept wonderfully free from colds, I believe by being more careful.

I have for some time had beside me the Answers to the Queries of the Education $Comm^n$ of the Med Council. I have not been in the humour to tackle them yet – I only hope there may be sufficient variety of authoritative opinion as will frustrate any proposal to make all medical teaching move at the same given rate on the same tread mill.

We shall have time to talk over these and other things when you come.
Mary unites with me know in kind regards —

  Yours always sincerely
   W Sharpey
Professor Allen Thomson
 &c.

---

[1] William Henry Colvill (1833–85), a medical officer in the army.
[2] John Storrar was the University of London's representative on the General Medical Council.
[3] Caesar Henry Hawkins (1798–1884), Sergeant-Surgeon to the Queen.

83

Hampstead London NW
6th Nov. 1870

My dear Thomson

I fear I can write you only very briefly – for I really have little or nothing worth telling you.

I am sorry I do not know Mr Redgrave [?] – but I have written to Seaton[1] who moves a great deal among official people, in the hope that he may be able to serve Mr Johnstone – of whom I think he must [have] been a fellow pupil.

I am glad to say that Sanderson[2] gets on well in our Practical Physiology Department. He has 34 pupils – not bad for a voluntary class with a rather high fee. I think moreover, that he will have two or three able young men to work at research in his Laboratory.

What we want in our Institution is *Money*. With four or five thousand pounds to lay out we could do wonders – but we have to work as well as we can with our small pecuniary means.

I read in yesterday's paper that there has been an unproarious demonstration in opposition to the new Professor of Midwifery in Edinburgh;[3] and that both Sir A Grant[4] & Christison were unable to procure him a hearing. Christison used to be very severe on untoward affairs of that sort which happened (exceptionally) in other schools; and used to vaunt the discipline of Edinburgh – but the peppering of Sir D. Brewster[5] with paper pellets which led for a time to the suppression of inaugural openings & the riot now described – ought to make him more considerate. It *is very difficult* for an academical Authority to deal with cases of this kind when the whole of the class are offenders – To shut up the theatre would give them what they fight for – & you have difficulty in making a fair selection for an example. It might answer perhaps to make every one answer to his name when called out by an Officer of the Univ[y] or run the risk of rustication – & the ringleaders might then be found out – but the task is very difficult when the fellows have a notion that they have good grounds for their opposition & think that justice should supersede discipline.

   Yours always sincerely
    W Sharpey

Dr Allen Thomson

[1] Presumably Edward Cator Seaton (1815–80), founder member of the Epidemiological Society.

[2] John Scott Burdon Sanderson (1828–1905) succeeded Michael Foster as Professor of Practical Physiology and Histology at University College London.

[3] I.e., Alexander Russell Simpson (1835–1916), nephew to the previous incumbent, J. Y. Simpson.

[4] Alexander Grant (1826–84), Principal of Edinburgh University since 1868.

[5] David Brewster (1781–1868), former Principal of Edinburgh University. Sharpey is mistaken in stating that Brewster was peppered with paper pellets; the Senate Minutes make it clear that he was bombarded with *peas*: Edinburgh University MSS, College Minutes, 1861–5, vol 2, p. 451.

84

Hampstead 16th Jany 1871

My dear Thomson

I am sorry I did not get your letter till I came home after post time tonight so that I could not write to you by return.

In Univ^y College there is a special Professor of Clinical Medicine and a special Professor of Clinical Surgery. Neither of these has any other chair in the College. Each has in-patients assigned to him in our Hospital.

Then the Professor of Medicine & the Professor of Surgery in the College have *virtute officii* in-patients in the Hospital and are also styled Professors of Clinical Medicine and Professors of Clin. Surgery. I don't know the rule at other Institutions in London – but as a matter of fact the Professors or Lecturers on Systematic Medicine and Surgery have also Hospital charges.

I notice from signs of grumbling (in the Brit. Med. Journ) because your students cannot conveniently get to the Infirmary to see Dispensary Patients,[1] and it occurs to me that you might at *once* open a place for *out*-Patients near the New College in attendant the full scheme.

I am glad to hear that John has definitely settled to be in London as his next move. I look on King's College as lucky in getting him.[2]

Please don't forget what I said about evidence concerning science from Glasgow – for the Royal Commission.[3] —

Yours always sinc^y

W Sharpey

Dr Allen Thomson

[1] After Glasgow University removed to Gilmorehill, and before the opening of the adjacent Western Infirmary, medical students were put to considerable inconvenience; they had to travel some two miles from the University to the Royal Infirmary for clinical instruction. See *Br. med. J.*, 1871, **i**: 41.

[2] I.e., Allen Thomson's son John Millar Thomson (1849–1933), who held a lectureship in chemistry at King's College London prior to his appointment as Professor.

[3] I.e., the Devonshire Commission on Scientific Instruction and the Advancement of Science, of which Sharpey was a member.

85

Wrote 2nd Feb. [This is in Thomson's hand.]

Hampstead 28th Jany 1872

My dear Thomson

I was glad to have your agreeable letter of the 24. Dec. – and I am now able to write to you the at least closely approximate time of meeting of the Gen Med. Council. viz the 27th of February, when of course we expect you as usual to be at Lawnbank – with Mrs Allen if she is disposed to move at all at this season.

I am glad to hear good accounts of your Class — The *new entries* at Univ^y College, which had risen for the last two years, have taken a great start this year – so that we head the list of New Students by a small majority. My new men (paying) are = 99 and I can make out the 100 by counting an "Epsom Scholar" who goes free. The dissection is so active that Ellis has had to place tables in the middle of our large room & increase the number at the sides. Mr Norman Macpherson is a prominent and estimable man, but he has got a capital bargain in Georgie.

The English scheme for common Examination has been formally agreed upon by the Colleges & virtually by the Universities.[1] It comes on for adoption by the U. of Londⁿ. next Wednesday – & Paget apprehends no real hindrance in the other Universities. Andrew Wood writes to Paget rather disappointingly as to *progress* in Scotland; and Alex. Wood (I should say) contemplates Parliamentary proceedings taking place & I believe he will get no one in this quarter willing to have recourse to Parliament until existing powers have been tried; and we hope that if the Med. Council sanctions our Conjunct Scheme under Section XIX. corresponding arrangements will be completed in no long time in Scotland & Ireland. These should not be stereotyped by Parliament until we have had experience of their working. Eventually, legislative authority may be needed. For one thing, the conjoint Certificate can only be made indispensable for registration for the present by the Licensing Bodies agreeing to withold their separate qualifications (for registration) until candidates have passed the Common Examⁿ but for permanent working it will be right to make the Common Examⁿ the sole adit to registration, & then the Degrees and Fellowships may be conferred – as it pleases the Authorities – on Candidates whether they have the Common qualification or not.

Paget tells me he is cognizant of what is doing in Ireland. He is very hopeful – for thirteen very sensible resolutions have been agreed to by the Delegates of the Authorities — I say of the Authorities – but Aquilla Smith,[2] who was in London on Thursday, speaks disappointingly of the Queen's University, which is completely under Corrigan's[3] thumb – & Sir Dominick is dissatisfied with the plan proposed – as he probably would be with any plan calculated to bring about a peaceful settlement.

I do not apprehend serious difficulties as to the imposition of further payments by Graduates. Should there be any difficulty at first it will speedily disappear – for the expenses of Medical Education dont keep pace with the increase of everything else & with the growing wealth of the Community. Moreover I dont think that restriction to a

Clinical Examination for men with degrees or fellowships would fulfil what is intended by a common qualification for admission to the Register.

The New Body will represent all the *Med*[1] Authorities – Universities as well as Corporations, and the purpose is to guarantee on behalf of the Public the fitness of Candidates irrespective of Examination by single licensing bodies. It has been agreed to accept the Univ[y] Exam. for the earlier branches of Study; as these are carried farther in the Universities than could be required by the Common Board & further examination upon them would harass Students to no purpose – but this is in so far a *concession* on the part of the *Conjoint* Authorities, who are not the *Corporations* alone, and *in principle* the Common controlling Examination would apply to all subjects of the professional Examination. We all here wish very much that a Scheme for Scotland may be ready for submission to the Gen[l] Council when it meets. If the three divisions of the Kingdom would agree to try each a plan for a term of years, the Med[l] Council would be very much relieved from further work until the time should come for confirmation (after amendment if needed) by Act of Parliament.

Yours always sincerely
W Sharpey

Dr A Thomson
  &c.

---

[1] In 1871 the Royal Colleges of Physicians and Surgeons of London framed a scheme for a Conjoint Examining Board which would qualify for medical practice in England. The GMC had in 1870 advised the licensing bodies of its wish to see such conjoint examining boards; and similar bodies were to be established in Scotland and Ireland.

[2] Aquilla Smith (1806–90), Professor of Materia Medica and Pharmacy at Trinity College, Dublin and representative of the Irish colleges on the GMC.

[3] Dominic John Corrigan (1802–80), Irish physician, MP for Dublin, and Vice-Chancellor of the Queen's University of Ireland. He opposed the idea of a joint examining board for Ireland.

86

6 Old Palace Gard
London 16th Feb. 1872
My dear Thomson

We have just had a very small meeting of the [Royal] Commission today; but it was understood by those present that you will be welcome at your own time – i.e. during the meeting of the Med[l] Council or immediately before. The 27th will be a blank day no doubt being set apart for the Thanksgiving. Wednesday the 28th would answer very well.

Mr. Sanderson has just come in after our meeting & he seems anxious for full information respecting Nat. Philos. and Chem. Laboratories of Glasgow University. I told him that Sir W. Thomson might come again if wanted for that object – but that I feared Dr T. Anderson's state of health would be an obstacle to his appearing – but I concluded by telling him that no one could give such good information respecting

Laboratories, Dissecting Rooms, Museums, Libraries &c. &c. than you – and, as what he chiefly wished for was *plans* of these several departments – that I would ask you to bring them up with you, or such copies as could be used by the Commission. I think he would wish them to be given in the printed minutes of the Commission.

I suppose that what the Commission would specially want from you would be information as to the applications of money in teaching science in the Univ^y – for Assistants, Apparatus, Collections &c. & what more may be wanted for effectually carrying on the work, the question of Assistant Professors in certain departments. Further development of influences & usefulness of the University as a scientific organization – e.g. towards instruction of school Teachers, practically and solidly in Science &c.

John [Millar Thomson] dined with us on Monday with Turner & [. . .] Brown[?]. He is strong and active. With kind regards from Mrs T

Yours always sincerely

W Sharpey

Dr Allen Thomson

87

Wrote 5th May [This is in Thomson's hand.]

Hampstead 30th April 1872

My dear Thomson

I have both your Letters. As regards Grant, I think the Commission would be glad to hear him if he has an inclination to come before them. He might give his opinion not only of the Glasgow Observatory & its relation to Observatories generally – and on the question of establishing new Observatories for observing the physical phenomena of the Sun and other celestial bodies. He may perhaps also have views as to general scientific Administration in connection with the Government.

Let me know about his wishes & I will look to the matter.

I am glad to get your proposals for a Conjoint Scheme. The difference from ours is in restricting the conjoint Exam^n to the subjects which now constitute the 2^d or final Examination and leaving the 1st Exam^n to be conducted singly by any of the present Licensing Bodies. I can see why the Corporations are averse to merging the whole of their Examinations into the Common one, and allowing an exception to Univ^y Graduates (as in our Scheme) and I suppose you will have to be contented with this in the mean time. Nevertheless it is "per tanto" a divergence from a concentrated examination, and to me appears a blemish. With that exception I think the proposed plan is excellent – and I venture to predict that after some experience the Corporations will find it more advantageous to combine all their work. Moreover I apprehend they will be disappointed if they expect that candidates who have passed the 1st Examination in Scotland (unless University Undergraduates) will be admitted to qualify under the joint scheme in England without going through the *whole* of the prescribed examination — [1]

It did not occur to me that there was much to amend in your Answers to the [Devonshire] Commission. But considerable license is allowed in correction – provided always that an answer shall not be so altered as to make the *question* inapplicable.

Yours always sincerely
W Sharpey

Dr Allen Thomson

---

[1] The GMC discussed plans for the creation of a conjoint examining board for Scotland at length in March 1872. See: *Br. med. J.*, 1872, **i**: 261–6. There were problems in following the English model because of the different relations between the universities and medical corporations in Scotland.

88

May 25 1872

I have been for some good while back troubled with incapacity of my bladder to keep a reasonable quantity, & consequently have frequent calls. This makes me anxious in travelling. I think you told me you had a contrivance which relieved you from all trouble. Will you kindly tell me what it is called – and where it may be got – also send a sketch of it & tell me how to use it.

M[ary?]. so far as I know is still in Arbroath

W.S.

89

Univ[y] College, London
24th October 1872

My dear Thomson

Mr William Longman[1] called here yesterday & I met him today by appointment in Paternoster Row. He wished to speak of Quain's Anatomy. His firm have bought the Copy right of it and he wished to advise us as to a new Edition.

I told him that when Rich[d.] Quain & I agreed to edit the 5th Edit. our bargain with the Publishers was £10..10 per sheet for new matter £3..3 for revision of old text. That when another (or a successive Editor) was wanted we were employed on the same terms as for the 5th (so long as we continued to be Professors of Anatomy & Physiology in the College[)]. Quain withdrew & Ellis took his place then I was left alone and an arrangement made with you & Cleland.

I explained that you & I had recommended a rearrangement of the subjects in a new Edition. One part to contain Histology & the structural & physiological anatomy of the viscera – the microscopic anatomy of the teeth – the organs of senses – and all development &c. the other part to contain the bare descriptive anatomy with the descriptions abridged. That with your approval and by desire of Walton I had urged Dr Gowers[2] to eliminate flaws [?] & condense the Second part & that he had been some time employed on it – & that whatever else was done I trusted he would receive suitable remuneration – this Mr L. promised – I then told him that my eyesight was so deteriorated (worse than when we parted) that I could scarcely work (in scientific reading or writing) at night – moreover that I could not examine objects critically with the microscope – and that I would not undertake what I had intended when we made our suggestions to Walton. Mr L. then asked me to recommend some one – but I decided first to communicate with you – indeed I told him you had the whole arrangement much better in your head than I have – & that after hearing from you I would speak with him again.

He wishes a distinct proposal to be made to his firm – & so far as I can see he will be quite ready to go on the same terms as Walton; but there is *no time to lose* – and even if done rather roughly – it would be *well* to have the book ready for next year's use. But who is to take my place? I dont like to press it on you as you would have enough to do independently. Could & would Cleland do – A good part of the matter might stand & additions & corrections put in when wanted. Michael Foster might do what is required – but I dont know what he may have on hand. But do you give your thoughts to the question & let me hear from you – especially put into writing your understanding as to the modifications in the new Edition.

I told Longman of the extra impression which Walton had run off (how many copies I could not say) & of the fact that his editor owed us a payment on that account. The Longmans have merely bought the copyright as offered in the market & know nothing of the extra impression, but Mr L. offered to make inquiry about it. I do not expect we shall ever get anything out of it.

It would be a great pity to let the book drop & if we cannot undertake the new Edition of course the Longmans may supply some other people – this I should greatly regret – The Book would now have a better chance than before.

I claim no right to be employed – on the contrary I am unable to accept the offer, but I see that the Longmans would be very much guided by my advice & I should be much guided *by you*.

Whatever else you write pray put a definite proposal on paper which I can extract & submit to the new publishers.

John spent a Sunday afternoon with me the other day – he is well & hopeful.

I think our new entries will be a *trifle* fewer this year – but the registration of the College of Surgeons places Univ[y] College and Guys *equal* in new entries & above the other schools – then we beat Guys in Second year men but they are higher in 3[d] and 4th a good lot of 4th year men with us did not think of registering.

With kind regards to Mrs Thomson. I have scarcely light to see
Your always most sincerely & attached
W Sharpey

Dr Allen Thomson
M.D. FRS
&c.

[1] William Longman (1813–77), partner in the publishing firm.
[2] William Richard Gowers (1845–1915), Assistant Physician at University College Hospital.

90

Nov 1873

My dear Thomson

I had occasion yesterday to confer with Sharpey and Burdon Sanderson on the subject of a proposal of a gentleman to endow the Chair of Physiology in University College with £300 a year.[1]

Sharpey took the opportunity of saying that his wish was to withdraw from his Chair at the end of the current session retaining the Directorship of the Museum which has a small salary attached to it. He said that his private means added to the salary of the Directorship would be enough to satisfy his wants. When Sanderson and I were alone he asked me if I knew enough of Sharpey's affairs to feel sure that he would really have a sufficiency to retire upon. My answer was that, I knew he had some savings that, his habits were not expensive and that, probably he had not overestimated his resources but that, I thought you might know more than I did of his private affairs and that I would write to you and inquire. Can you give me any information on this head? As a member of the College Council I can say that no Professor is more respected by them and that they would one and all be anxious to make the remainder of Sharpey's days easy but on the other hand I must acknowledge that the resources at our disposal and the precedent in relation to others would make it difficult to [*sic*] us to assign a pension to him. It would be a great relief to me and Sanderson, and I am sure to others also, to feel sure he did not need it but in case of uncertainty we should like to consider what it might be practicable to do for him.

As regards retirement I should not like for Sharpey's sake to dissuade him from taking a step which will be honourable to him. At the end of this Session he could withdraw with undimmed reputation; it would be a risk for him to hang on longer except as a means of introduction to a successor by giving a few lectures.

At your convenience will you give me a few lines. Sharpey is in fair health and spirits but his sight is very defective and I doubt if it is likely to mend.

You will be surprised to see we have moved into town. Various reasons have reconciled us to leaving Hampstead and we are very comfortable here.

My wife is pretty well and unites with me in kindest regards to you and Mrs Thomson.

Believe me
Yours most faithfully
John Storrar

Let Sharpey's intentions be a secret for the present.

---

[1] I.e., the bequest which led to the creation of the Jodrell Chair of Physiology at University College.

91

University, Glasgow
11th November 1873

My dear Dr. Quain,

I need not attempt to express the pleasure I received from the assurance conveyed by your letter of yesterday that the proposal to acknowledge our friends long and faithful services in the Cause of education and scientific progress by some substantial reward was likely to be well received in influential quarters.

Nor can I tell you how grateful I feel to you for your prompt exertions in the cause. I wish sincerely I could in any way assist; but I fear I have no influence, beyond the strong expression of the opinion of one who, having been on terms of the most intimate friendship with Sharpey for five and forty years is perhaps better acquainted with his deserts than any other person living. In our frequent and often prolonged intercourse through the whole of that period I have ever felt the influence of his elevated character and sound judgment in benefitting myself. I know this feeling to be shared by all those who have had the happiness of his friendship as well as by all his pupils — I know too the sacrifice he made in declining the Edinburgh Chair of Anatomy, in the possession of which he would have been wealthy beyond most scientific men, and from which he could now have retired upon more than a competency. But beyond all this all the world knows with what singular devotion, ability, judgment and entire negation of self Sharpey has spent the whole of his life in the pursuit and advancement not merely of his own special department but of a considerable range of sciences — His retiring disposition alone has prevented him from occupying more prominent positions; And I cannot but feel that while his present circumstances call urgently for something being done to smooth the course of his advanced years, the Government in giving him an ample pension would not only do an act of justice, certain to meet with universal approval, but w^d also do itself honour in marking its appreciation of such faithful and able exertions and of so disinterested a life.

I need only add that I know that without provision from other sources than his own property it would be impossible for Sharpey to live at all comfortable after a retirement from his present office.

Earnestly hoping that you may be successful in your efforts,

I am ever sincerely yours

AT

[This is a heavily corrected draft of the letter.]

92

Wrote 25th [This is in Thomson's hand.]

50 Torrington Square, Lond.
24th Feb. 1874

My dear Thomson

I have returned the specimens to Mr Longman, preferring N$^{o.}$ 2.

I meant to have written yesterday, as in duty bound to my old and attached friend, to

135

inform him that on Saturday I had a most kind letter from Lord Granville containing the most unlooked for intelligence that I had been assigned a pension of £150 on the Civil List. I suspect that you and Dr. Quain must have first moved in the matter.[1]

As it is this very welcome resource added to my stipend here (of £100) which I expect to retain on retiring, as I shall do from my *Professorship* at the end of this session – together with what I shall have in other ways, will put me *quite* at my ease.

I may tell you as my friend, and privately, that I have very nearly £200 a year from investments and hope to make it more by investing part of what is now at my Banker's – what [*sic*] ought to bring £50. Then my nephew W^m. H^n. Colvill, who was educated & brought up as well as set out at my charge – makes me a very liberal contribution of £200 a year.

This he could not afford to do were he to retire from the Service, which must some time come to pass – but even in that case I should, as you see, be very well off.

Parkes[2] was here examining for the Army on Saturday. He spoke favourably of John's progress towards recovery.

As to my eyesight, I am alternately despondent and hopeful. I gave along account of my state to Reid[3] – saying he might show it to you. He has written me his opinion which I regard as very important. This letter is written with my "operated eye" with the aid of glasses combining a cylindrical with a spherical curve.

Your acc^t of Mrs W^m4 is very serious —

Your sincere friend

W Sharpey

[1] Sharpey's pension was announced in the *Br. med. J.*, 1874, i: 354–5.
[2] Edward Alexander Parkes (1819–76), Professor of Hygiene at the Army Medical School, Netley.
[3] Presumably Thomas Reid (1830–1911), Surgeon at the Glasgow Eye Infirmary and Lecturer on Ophthalmic Medicine at the University of Glasgow.
[4] Presumably William Thomson's widow.

93

London, 16th Decr 1875

Mr dear Thomson

A. B[uchanan].'s printed letter is indeed a (deplorable) curiosity.[1] It is not merely dotage bursting out under temporary irritation – it shows an originally ill cultivated mind uncorrected by culture. It is monstrous to reflect that such a *common* kind of man should have been your Professor.

I will write to Foster about Balfour[2] if you should still wish it – but I doubt if it would be advisable to look that way. The Physiology of [*sic*] must turn chiefly on function and its investigation through physics and chemistry – it must also take in histology — Now although I don't know that Balfour might not take ably and with relish to that side of the Subject, we know that his taste and favourite work have hitherto been in Morphology and Embryology — Your man must also be a good Human Anatomist as distinct from a Zootomist.

Moreover, I scarcely think the plan would tempt Balfour – He has independent income of his own. As Fellow of Trinity I suppose he has £350 to £400 a year besides his rooms and dinners in Hall. Were I he I should keep myself free – to go to Naples, Norway or America when I liked.

I am glad that McKendrick[3] will apply – he is very much the man you want – up to modern physiology in a systematic way and long-versed in teaching Experimental Physiology – I should think he would be a great acquisition. I have heard of Mr. Stirling[4] most favourably – but beyond this I could form no opinion.

But then your supreme difficulty is what to do for the remainder of this session! You cannot leave the students utterly without instruction.

I hope you will not be worried with this troublesome business more than you can help. I fear Andrew [Buchanan] will make a fight about allowing a substitute to appear in his theatre – it might show that his difficulty of making himself heard is not solely due to acoustic defects in its construction.

As things are now – perhaps it is an advantage that he cannot be heard —

Yours most sincerely

W Sharpey

In haste

---

[1] In May 1875 Andrew Buchanan, Professor of the Institutes of Medicine at Glasgow, had published a sharp rejoinder to critical remarks about his teaching and standard of examining, made by members of the GMC. He appended a list of examination questions to his letter, challenging his detractors to submit answers to the *Lancet*. See the *Lancet*, 1875, **i**: 661. In December 1875 Buchanan announced his retirement from the chair.

[2] Francis Maitland Balfour (1851–82), embryologist. He was appointed lecturer on animal morphology at Cambridge in 1875.

[3] John Grey McKendrick (1841–1925), lecturer on physiology in the Edinburgh extramural school. He succeeded Buchanan as Professor of Physiology at Glasgow in 1876.

[4] Presumably William Stirling (1851–1932), assistant to the Professors of Natural History and Physiology in Edinburgh. He became Professor of Physiology at Aberdeen in 1877 (see letter 99).

94

London 28. Dec 1875

My dear Thomson

I return by Book Packet [?] your two Sheets of Embryology. Do not bother yourself with my pencillings unless they consist of a positive mistake.

The book is promised in January – and considering the delay & disappointments as well as the expedient had recourse to of separate previous issue of Vol 1, it is absolutely indispensable that Messrs Longman should be able to keep promises. I beg therefore that you will not allow yourself to be distracted or diverted from the work by *any other* occupation except the delivery of your Lectures.

Foster happened to call on me shortly after I wrote to you last, I found he entirely agreed with me as to the unsuitability of the Gl. Chair for Balfour. He would take McKendrick.

It would be most unfortunate if local influence should be allowed to prevail.[1]
    Yours always sincerely
     W Sharpey
       nearly in the dark
Dr A Thomson

[1] McKendrick's claims to the Physiology Chair were challenged by a local candidate, Eben Watson (1824–86), Professor of Physiology at the Andersonian College (1850–76) and Surgeon at the Glasgow Royal Infirmary.

95

        50 Torrington Square Monday mor[n]
          (21st Aug[t] 1876)
My dear Thomson
   I inclose a letter I have just received from Jenner.[1]
   Sanderson has forwarded to me the joint representation[2] desiring me to sign it & send it to Hooker[3] (whose address is "care of F. Symonds Esq. Bridge Street Hereford["]).
   Sanderson called on Paget on Saturday but found he had gone to Switzerland. If his name cannot be got I think (unless Acland joins) that it would be better for Hooker and Sanderson to write to Cross[4] separately, especially as I doubt whether my name should be added (seeing that I have virtually written to Mr. Cross separately[)]. I have indeed signed the paper but I think that both Sanderson & you will rather approve of its not appearing & I have suggested to Sanderson to write out a fresh copy for Hooker's name should it still be thought advisable to send a joint recommendation.
   I have returned the paper to Sanderson for the reason mentioned and also because if he wants his own signature –
Hoping I may see you tomorrow
    Yours always
     W Sharpey
Dr Thomson FRS

P.S.   I have told Sanderson of your movements – and I shall be all day either here or at the College.
           WS.

[1] William Jenner (1815–98), Physician-in-Ordinary to the Queen.
[2] See letter 98.
[3] Presumably the botanist Joseph Dalton Hooker (1817–1911).
[4] Richard Assherton Cross (1823–1914) Home Secretary in Disraeli's government from 1874 to 1880. The Glasgow physiology chair was in the gift of the Crown.

96

Balmoral Castle

August 19th
1876

My dear Sharpey

As I shall I hope be away from Balmoral before Mr. Cross arrives (*i.e.* late in Sep^t I am told) I thought it better to send your letter to him at once – explaining to him the high value to be set on your Testimony – and I trust Dr. McKendrick may obtain the post – I have said & done what I think best calculated to aid him.

Yours always very Sinc^y
W. Jenner

97

Wrote in March & on 15th April [This is in Thomson's hand.]
London 17th Feb. 1877

My dear Thomson

I had the [Glasgow] "Herald" this morning, with the beginning and end of your Lecture.[1] I have no doubt your exposition of the mode of development & its relation to the structure in animals must have been lucid and instructive. Your conclusions as to the dependence of mental manifestations on the cerebral organization are doubtless incontrovertible if your Darwinism (gently hinted) may be right, but you dared not have said as much when you suceeded Alison in Edinburgh & if you had said as much before, the municipal Patrons (who would have refused Agassiz[2] for claiming more parents for the human race than Adam & Eve) would never have promoted you! One is glad to find so liberal an Audience & enlightened a Chairman as you enjoyed in Glasgow.

Now for myself. The first trials of my eye after operation were very promising – but a film soon occupied the visual opening – whether by deposit on the post^r lens capsule or independent of it I cannot tell. The appearance of the opening seen by me by aid of a pinhole – you may guess from the rude sketch I inclose viz. A. [see figure 8] The film persisting and rather getting worse, Streatfield with Couper's[3] approval and assistance endeavoured to split it up with a needle. They had hoped it would spring open to the needle-cut. But it was tough and unmanageable and all that could be effected was to peel it away to the temporal side – about half way across – leaving one half floating in the aqueous humour with a thick border running vertically & leaving about half the area clear to the nasal side. The membrane quivers with the movements of the eyeball but is still fixed somewhere on the temporal side.

The natural pupil was occupied with a *dense* looking brown mass – which is now half turned away. The sketches are placed as they appear to my vision by homocentric light – but as you know they are inverted and you must turn them to get the actual position.

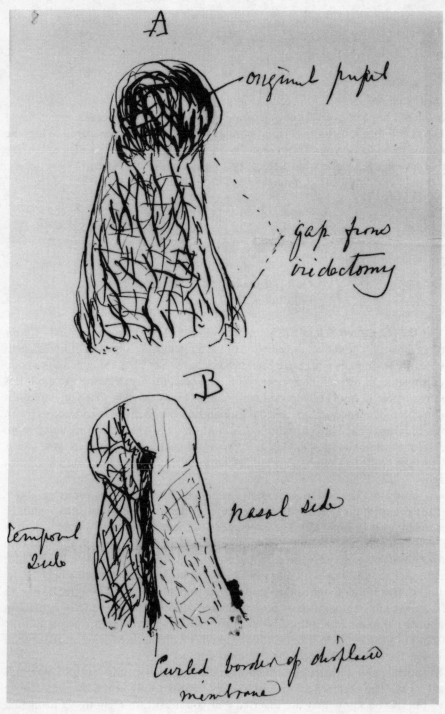

Figure 8.

Meanwhile I have now had spectacles for a week, made under J. Couper's direction – with a cylindrical error of 15/inches radius — The distance glasses have hitherto answered well, but I am not satisfied with the near ones – or rather I ought to say that the obstruction in the eye tells more against "reading vision."

It is just possible that the offending membrane may yet shrink up – or become wasted, but I have no great hope of this, and I am contented to go on in the mean time in thankfulness especially as I can now walk the streets without fear.

Pray ask Reid kindly to say what he thinks[.] I hope Mr Macilwain's operation is now showing some result.

Mrs Sibson called two days ago to consult about the best depositing for her late husband's[4] MSS, Drawings &c. – and she brought with her a framed print of your Father which I had given to Sibson some time ago. She is now giving up the house in Brook Street & has no space in the country (where she will now live) for hanging pictures – so she thought it best to return the one in question. Do you know any one who would like to have it? Had I not been writing to you I should have asked John – whether Mrs Aiken would care for it – but perhaps you may suggest a suitable bestowal of it.

John kindly assented to be one of my Trustees & I have accordingly made a new Will and named Marshall & John as Trustees.

I have heard from Wm Henry [Colvill] – first from Constantinople & this morning from Cairo. He has got the Sanitary Commission at Constantinople to recommend two measures proposed by him in reference to quarantine and plague – I dont suppose he will do anything in Egypt which he visits merely as on the route to Bombay, whence he will travel by Steamer to Busrah [*sic*] & Bagdad.

> With kind regards to M^{rs} T.
> Yours always sincere friend
> W Sharpey

Dr A. Thomson FRS

---

[1] Thomson had given the last of a series of lectures on 'The evolution of the brain' at the City Hall, Glasgow. See the *Glasgow Herald*, 16 February 1877.

[2] Jean Louis Rodolphe Agassiz (1807–73), Swiss naturalist who strongly opposed the Darwinian theory.

[3] John Fremlyn Streatfield (1828–86), and John Couper (1835–1918), Surgeons at the Royal London Ophthalmic Hospital. These remarks refer to an eye operation Sharpey had undergone in October 1876.

[4] I.e., Francis Sibson (1814–76), who had studied medicine at Edinburgh before becoming Physician to St Mary's Hospital, London.

98

London 50 Torrington Square
3^d Sep^t 1877

My dear Thomson

The case is not an easy one to deal with.[1] When the Physiology Chair was in question, the competitor of McKendrick had brought to bear very strong local influence – and this publicly and avowedly in his favour, as a Glasgow Man – and there was no difficulty in joining in a remonstrance against the possible sacrifice of the interest of the University and the scientific reputation of the Chair to local clamour and political pressure – but in the present case any representation to Mr. Cross could scarcely refer openly to Court influence – as opposed to scientific merit. No doubt a strong expression in favour of bestowing the place on a Candidate who would maintain & enhance the reputation of the institution – might properly be made – but as there are two candidates who fulfil that condition – the interposition proposed – would simply be a warning against the third; and then his tried success as a teacher who has worked on in Glasgow in view of getting into the Univ^y might be urged in his favour.[2] I confess I am at a loss what to advise — I have seen Sanderson this morning – he feels that his position scarcely warrants him in taking a prominent part as he did when *physiology* was the question – but I doubt not he would join in any measure which should be supported by any considerable number of scientific men — This however can scarcely come about. We both think you might write to Huxley and to Owen and learn what they could or would do — Also Turner[3] & Struthers[4] might aid – tho' perhaps Turner might not care to see a powerful rival in Glasgow. Why not put the case before your Chancellor Sir. W. Maxwell?[5] If you would move in it the Gov^t would be strengthened in resisting Court influence – supposing Mr. Cross wished to resist it. The chief use of a round-robin would be to create an outside cry (in the Medical Profession) against throwing away the opportunities of promoting science.

I called at Busk's house & find he is at Ambleside – his address is "Hill-top" or "Top-Hill" I forget which, Ambleside.

I wrote some time ago to Jenner in favour of Stirling for Aberdeen & got a most friendly answer, but Jenner could only communicate wth Mr. Cross if they should meet – for when he wrote to him about the Edin. Clin. Surg^y Chair[6] he was told by Mr. C's Secretary substantially but in civil terms that he had better wait till his opinion was asked.

You will gather from this disjointed note that I feel quite at Sea in the matter.

Suppose you begin by drawing up a short address to Mr. Cross & send it for consideration to such people as you would like to join in it – say Owen, Huxley, Busk[,] Sanderson[,] Flower[,][7] Turner[,] Struthers & me. I cannot find whether Huxley is in London. But write to the usual address (to be forwarded).

Yours sincerely
W Sharpey

Dr Allen Thomson

---

[1] Thomson vacated the Anatomy Chair at Glasgow in 1877; he was succeeded by his former assistant John Cleland.

[2] This third candidate may have been George Buchanan, a lecturer on anatomy at the Andersonian College. On the second "desirable" candidate see the next letter.

[3] William Turner (1832–1916), Professor of Anatomy in Edinburgh.

[4] John Struthers (1823–99), Professor of Anatomy in Aberdeen.

[5] William Stirling-Maxwell (1818–78) landed magnate and historian. He had been Chancellor of Glasgow University since 1876.

[6] The Edinburgh Clinical Surgical Chair became vacant in 1877 upon Lister's transfer to King's College, London.

[7] William Henry Flower (1831–1899), Hunterian Professor of Comparative Anatomy at the Royal College of Surgeons of London.

99

50 Torrington Square London WC
22^d Sep^t 1877

My dear Thomson

I have within the last two or three days been trying a little practical work in the College – looking over my old preparations of the bitch's Uterus.

I was seized yesterday with my vertigo – after being free from it for nearly four months – but I hope to be right again – and should be glad to have a specimen of the unimpregnated uterus of the bitch also of an impregnated one in which the foetal compartments are about the size of a walnut or a little less. It occurs to me that you might fall on something of the kind in rummaging among your old traps during the process of their removal to the Univ^y Museum – and I therefore ask you kindly to help me.

I saw Stirling on Wednesday – on his way to Paris to look after teaching and working appliances suited to his new position [as Professor of Physiology at Aberdeen] – his attainment of which I need hardly say has been a great satisfaction to friends here – not only on his account but as some token that Government patronage in the Scottish Universities will be thoughtfully and judiciously bestowed. He tells me that your *succesion* will be between Cleland and Macalister[1] – and I hope he may be right. It would help to a satisfactory conclusion if some preference were come to between the two desirable candidates. Mr. Cross I see by the papers is at Balmoral and Jenner is there – Should Jenner be consulted I have no doubt he will give good advice. On all hands, that is, among all people who set the interests of Science above personal favour – it is anxiously hoped that so important a Chair will be worthily filled.

The St. George's people have I suspect made a mistake. Wanklyn[2] is no doubt a clever man and of considerable *scientific* reputation – but he does not get on with brother chemists. John [Millar Thomson] would have been a much better man for them.

Hoping to see you in no long time
Yours always as before
W Sharpey
Dr Allen Thomson

---

[1] Presumably Alexander Macalister (1844–1919), Professor of Comparative Anatomy and Zoology at the University of Dublin.
[2] James Alfred Wanklyn (1834–1906), Lecturer in Chemistry and Physics at St George's Hospital, London.

Appendix

Calendar of Sharpey-Thomson letters and associated correspondence in Glasgow University Library (MS Gen. 1476). The numbers of the letters reproduced here are indicated.

BOX 15: *Sharpey-Thomson letters*

Letter 1:	15 July 1836	
	Thomson	to Sharpey
		3 Aloa St [Edinburgh]

Letter 2:	18 July 1836	
	Sharpey	to Thomson
	Edin[burgh]	care of
		John Murray Esq.
		50 Albermarle Street
		London

Letter 4:	(1836)	
	Sharpey	to Thomson
	(London)	
	[Date and address in Thomson's hand.]	

Letter 5:	3 September 1836	
	Sharpey	to Thomson
	Edin[burgh]	(The Doune of Rothiemurchar)
	[Notes of Thomson's reply to Sharpey on the back of this letter.]	

Letter 6:	6 October 1836	
	Thomson	to Sharpey
	The Doune	
	Lynwilg	
	(by Perth)	

Letter 7:	2 December 1836	
	Sharpey	to Thomson
	25 Dover St	
	London	

Letter 8:	24 December [1836]	
	Thomson	to Sharpey
	[Notes written on the wrapper of Sharpey to Thomson, 2 December 1836.]	

Letter 9:	25 February (1837)?	
	Sharpey	to Thomson
	25 Dover St [London]	
	[Year and question mark in Thomson's hand.]	

	24 March 1837	
	Sharpey	to Thomson
	25 Dover St [London]	

Letter 10:	17 May 1837?
	Sharpey	to Thomson
	25 Dover St [London]
	Dr. Willis's?
	[Date and address in Thomson's hand.]

Letter 11: 8 December 1837
Sharpey   to Thomson
University College [London]

Letter 12: 30 April 1838
Sharpey   to Thomson
68 Torrington Sq [London]   at his Grace the Duke of
                 Bedfords

[This is coupled with a letter of 11 March (1838) from]

Margaret [Millar?]   to Thomson
10 Paris Place
Great Ormond St [London]

Letter 13: 24 November [1838]
Sharpey   to Thomson
68 Torrington Sq [London]

Letter 14: 16 February 1839
Sharpey   to Thomson
London

Letter 15: [February? 1839]
Thomson   to Sharpey

Letter 16: 9 March 1839
Sharpey   to Thomson
London   80 George St
                 Edinburgh

Letter 17: 4 December 1839
Sharpey   to Thomson
68 Torrington Sq
London

Letter 18: 28 November 1841
Sharpey   to Thomson
68 Torrington Sq
London

27 May 1842
Sharpey   to Thomson
London

Letter 19: 27 June 1842
Sharpey   to Thomson
68 Torrington Sq
London

Letter 20:	10 August 1842 Sharpey London	to Thomson
Letter 21:	13 October 1842 Sharpey London	to Thomson
	6 June 1844 Sharpey London	to Thomson
Letter 22:	4 May 1845 Sharpey 35 Gloucester Crescent London	to Thomson
Letter 23:	11 June 1845 Sharpey 35 Gloucester Crescent [London]	to Thomson
Letter 24:	29 October 1845 Sharpey London	to Thomson
Letter 25:	27 December 1845 Sharpey 35 Gloucester Crescent London	to Thomson
Letter 29:	9 March 1846 Sharpey London	to Thomson
	2 April 1846 Sharpey	to Thomson
	19 October 1846 Sharpey London	to Thomson
Letter 30:	23 December 1846 Sharpey 35 Gloucester Crescent London	to Thomson
Letter 32:	16 March 1847 Sharpey 35 Gloucester Crescent London	to Thomson
	1 April 1847 Sharpey London	to Thomson

	5 April 1847 Thomson 10 Hope St Edinburgh	to Sharpey
Letter 33:	8 April 1847 Sharpey London	to Thomson
	2 August 1847 Sharpey London	to Thomson
Letter 34:	20 December 1847 Sharpey London	to Thomson
Letter 35:	28 December 1847 Sharpey 35 Gloucester Crescent London	to Thomson
Letter 36:	9 January 1848 Sharpey 35 Gloucester Crescent [London]	to Thomson
Letter 37:	25 January 1848 Sharpey London	to Thomson
Letter 38:	[26–31 January 1848] Sharpey London	to Thomson
Letter 39:	31 January 1848 Sharpey London	to Thomson
Letter 40:	8 February 1848 Sharpey London	to Thomson
Letter 41:	14 February 1848 Sharpey London	to Thomson
Letter 42:	18 May 1848 Sharpey London	to Thomson
Letter 43:	31 August 1848 Sharpey London	to Thomson

Letter 44:	4 January 1849 Sharpey 35 Gloucester Crescent London	to Thomson
Letter 45:	7 March 1849 Sharpey 35 Gloucester Crescent [London]	to Thomson
	18 June 1849 Sharpey Littlehampton Sussex	to Thomson
Letter 46:	9 December 1849 Sharpey London	to Thomson
	12 July 1850 Sharpey Arbroath	to Thomson
Letter 47:	14 January 1851 Sharpey London	to Thomson
	15 April 1851 Sharpey London	to Thomson
	17 May 1852 Sharpey London	to Thomson
Letter 48:	19 July 1852 Thomson College Glasgow	to Sharpey
Letter 49:	20 July 1852 Sharpey London	to Thomson
	4 February 1853 Sharpey 35 Gloucester Crescent London	to Thomson
	28 February 1853 Sharpey London	to Thomson
	16 August 1853 Sharpey London	to Thomson

Letter 50:	19 May 1854 Sharpey 35 Gloucester Crescent London	to Thomson
	25[?] June 1854 Sharpey London	to Thomson
	11 April 1855 Sharpey London	to Thomson
	13 April 1855 Sharpey London	to Thomson
	29 April 1855 Thomson Greenhall	to Sharpey
Letter 51:	15 August 1855 Sharpey London	to Thomson
Letter 52:	2 September 1855 Sharpey Ballater Deeside	to Thomson
Letter 53:	15 October 1855 Sharpey 33 Woburn Place London	to Thomson
Letter 54:	27 October 1855 Sharpey 33 Woburn Place London	to Thomson
Letter 55:	5 December 1855 Sharpey 33 Woburn Place London	to Thomson
Letter 56:	9 December 1855 Thomson College Glasgow	to Sharpey
	17 December 1855 Sharpey 33 Woburn Place London	to Thomson

	30 December 1855 Sharpey Hastings	to Thomson
	13 January 1856 Sharpey	to Thomson
	9 March 1856 Sharpey 33 Woburn Place London	to Thomson
	4 May 1856 Sharpey 33 Woburn Place London	to Thomson
	1 February 1857 Sharpey 33 Woburn Place London W.C.	to Thomson
Letter 57:	14 March 1857 Sharpey 33 Woburn Place London W.C.	to Thomson
	6 April 1857 Sharpey 33 Woburn Place London W.C.	to Thomson
	9 May 1857 Sharpey The Royal Society Somerset House [London]	to Thomson
Letter 58:	10 June 1857 Sharpey 33 Woburn Place London W.C.	to Thomson
	10 July 1857 Thomson Hatton House Ratho	to Sharpey
	14 July 1857 Sharpey 33 Woburn Place [London] W.C.	to Thomson
	28 October 1857 Sharpey 33 Woburn Place London W.C.	to Thomson

Letter 59:	15 November 1857 Sharpey 33 Woburn Place London	to Thomson
Letter 60:	21 November 1857 Sharpey 33 Woburn Place London W.C.	to Thomson
	28 February 1858 Sharpey 33 Woburn Place [London]	to Thomson
	23 April 1858 Sharpey 33 Woburn Place London W.C.	to Thomson
	28 April 1858 Sharpey 33 Woburn Place London	to Thomson
	15 May 1858 Sharpey The Royal Society Burlington House [London]	to Thomson
Letter 61:	10 August 1858 Thomson Ratho House Ratho	to Sharpey
Letter 62:	12 August 1858 Sharpey University of London Examination	to Thomson
Letter 63:	26 November 1859 Thomson The College Glasgow	to Sharpey
Letter 64:	27 November 1859 Thomson The College Glasgow	to Sharpey
Letter 66:	2 January 1860 Thomson The College Glasgow [Written in Edinburgh.]	to Sharpey

Letter 65:	2 January 1860 Thomson Edinburgh	to Sharpey
Letter 67:	6 January 1860 Thomson The College Glasgow	to Sharpey
Letter 69:	18 January 1860 Sharpey London	to Thomson
Letter 70:	20 January 1860 Sharpey 33 Woburn Place [London]	to Thomson
	6 August 1860 Sharpey Burlington House [London]	to Thomson
Letter 71:	16 December 1860 Thomson College Glasgow	to Sharpey
	7 March 1861 Sharpey 33 Woburn Place London W.C.	to Thomson
	11 April 1861 Sharpey 33 Woburn Place London W.C.	to Thomson
Letter 72:	4 March 1863 Sharpey 33 Woburn Place London W.C.	to Thomson
	9 March 1863 Sharpey The Royal Society Burlington House [London]	to Thomson
Letter 73:	10 March 1863 Thomson College Glasgow	to Sharpey
	15 May 1863 Sharpey 33 Woburn Place [London] W.C.	to Thomson

	21 December 1863 Thomson College Glasgow	to Sharpey
Letter 74:	31 December 1863 Sharpey 33 Woburn Place [London] W.C.	to Thomson
Letter 75:	12 July 1864 Sharpey 33 Woburn Place [London]	to Thomson
	10 September 1864 Sharpey 33 Woburn Place London W.C.	to Thomson
Letter 76:	25 October 1864 Sharpey 33 Woburn Place London W.C.	to Thomson
	29 December 1864 Sharpey 33 Woburn Place London	to Thomson
Letter 77:	18 February 1865 Sharpey 33 Woburn Place London	to Thomson
Letter 78:	10 December 1865 Sharpey 33 Woburn Place [London] W.C.	to Thomson
	30 December 1865 Sharpey 33 Woburn Place London W.C.	to Thomson
	11 June 1866 Sharpey 33 Woburn Place [London] W.C.	to Thomson
	27 October 1866 Sharpey 33 Woburn Place London W.C.	to Thomson

Letter 80:   2 January 1867
Sharpey   to Thomson
33 Woburn Place
London

21 January 1867
Sharpey   to Thomson
33 Woburn Place
London W.C.

29 January 1867
Sharpey   to Thomson
University College
London

27 January 1868
Sharpey   to Thomson
Lawnbank
Hampstead
[London] N.W.

Letter 81:   15 November 1868
Sharpey   to Thomson
Hampstead
London N.W.

Letter 82:   29 March 1869
Sharpey   to Thomson
Hampstead
London N.W.

15 June 1869
Sharpey   to Thomson
Hampstead
London N.W.

19 June 1869
Sharpey   to Thomson
Lawnbank
Hampstead
London N.W.

4 January 1870
Sharpey   to Thomson
Hampstead [London]

Letter 83:   6 November 1870
Sharpey   to Thomson
Hampstead
London N.W.

7 November 1870
Sharpey   to Thomson
Hampstead [London]

Letter 84: 16 January 1871
Sharpey   to Thomson
Hampstead [London]

13 June 1871
Sharpey   to Thomson
London

26 June 1871
Sharpey   to Thomson
London

Letter 85: 28 January 1872
Sharpey   to Thomson
Hampstead [London]

Letter 86: 16 February 1872
Sharpey   to Thomson
6 Old Palace Gardens
London

28 April 1872
Sharpey   to Thomson
Hampstead [London]

Letter 87: 30 April 1872
Sharpey   to Thomson
Hampstead [London]

7 May 1872
Sharpey   to Thomson
Hampstead [London]

Letter 88: 25 May 1872
Sharpey   to Thomson
Hampstead [London]

26 May 1872
Thomson   to Sharpey
Morland

8 June 1872
Sharpey   to Thomson
Hampstead [London]

9 July 1872
Sharpey   to Thomson
Hampstead [London]

14 July 1872
Sharpey   to Thomson
Hampstead [London]

	19 July 1872 Sharpey Broughty Ferry	to Thomson
	15 August 1872 Sharpey Wildbach	to Thomson
	9 September 1872 Sharpey Klobenstein	to Thomson
	25 September 1872 Sharpey Hampstead [London]	to Thomson
Letter 89:	24 October 1872 Sharpey University College London	to Thomson
	25 December 1872 Sharpey 50 Torrington Square London W.C.	to Thomson
	25 January 1873 Sharpey 50 Torrington Square London W.C.	to Thomson
	12 March 1873 Sharpey 50 Torrington Square London W.C.	to Thomson
	22 April 1873 Sharpey 6 Old Palace Gardens London	to Thomson
	9 May 1873 Sharpey University College London	to Thomson
	16 August 1873 Sharpey 50 Torrington Square London W.C.	to Thomson
	10 September 1873 Sharpey Embden House Broughty Ferry	to Thomson

28 September 1873
Sharpey   to Thomson
50 Torrington Sq
London

14 November 1873
Sharpey   to Thomson
London
[There are notes on this letter by Thomson regarding provision for the
Hunterian Museum and Botanic Gardens in Glasgow.]

Letter 92:   24 February 1874
Sharpey   to Thomson
50 Torrington Sq
London

12 May 1874
Sharpey   to Thomson
Torrington Sq [London]

23 June 1874
Sharpey   to Thomson
50 Torrington Sq
London

24 June 1874
Sharpey   to Thomson
50 Torrington Sq
London

4 September 1874
Sharpey   to Thomson
University College
London

13 September 1874
Sharpey   to Thomson
Dr. Arrott's
Arbroath

28 November 1874
Sharpey   to Thomson
London

29 January 1875
Sharpey   to Thomson
50 Torrington Sq [London]

17 February 1875
Sharpey   to Thomson
Athenaeum Club
London

20 March 1875
Sharpey   to Thomson
50 Torrington Sq [London]

26 March 1875
Sharpey   to Thomson
50 Torrington Sq [London]

29 July 1875
Sharpey   to Thomson
50 Torrington Sq [London]

16 August 1875
Sharpey   to Thomson
50 Torrington Sq [London]

19 August 1875
Sharpey   to Thomson
50 Torrington Sq [London]

25 August 1875
Sharpey   to Thomson
50 Torrington Sq [London]

5 September 1875
Sharpey   to Thomson
Grant Arms
Grantown
Invernesshire

13 September 1875
Sharpey   to Thomson
Grantown

16 September 1875
Sharpey   to Thomson
Grantown

24 September 1875
Sharpey   to Thomson
Kinloch House
Blairgowrie

29 September 1875
Sharpey   to Thomson
Arbroath
[Date in Thomson's hand.]

8 November 1875
Sharpey   to Thomson
London

10 November 1875
Sharpey   to Thomson
London

	13 November 1875 Sharpey London	to Thomson
	23 November 1875 Sharpey 50 Torrington Sq London	to Thomson
	26 November 1875 Sharpey London	to Thomson
	3 December 1875 Sharpey London	to Thomson
Letter 93:	16 December 1875 Sharpey London	to Thomson
Letter 94:	28 December 1875 Sharpey London	to Thomson
	7 February 1876 Sharpey University College London	to Thomson
	11 February 1876 Sharpey 50 Torrington Square London	to Thomson
	18 February 1876 Sharpey 50 Torrington Square London	to Thomson
	7 March 1876 Sharpey University College London	to Thomson
	3 April 1876 Sharpey 50 Torrington Square London	to Thomson
	4 April 1876 Sharpey 50 Torrington Square London	to Thomson

	7 April 1876 Sharpey 50 Torrington Square London	to Thomson
	9 April 1876 Thomson University Glasgow	to Sharpey
	5 July 1876 Sharpey Arundel Sussex	to Thomson
	7 August Sharpey 50 Torrington Square [London] W.C.	to Thomson
Letter 95:	21 August 1876 Sharpey 50 Torrington Square [London]	to Thomson
	13 October 1876 Sharpey 50 Torrington Square [London]	to Thomson
	December 1876 Sharpey 50 Torrington Square [London]	to Thomson
	16 April 1877 Sharpey London	to Thomson
Letter 97:	17 February 1877 Sharpey London	to Thomson
Letter 98:	3 September 1877 Sharpey 50 Torrington Square London	to Thomson
Letter 99:	22 September 1877 Sharpey 50 Torrington Square London W.C.	to Thomson
	16 December 1877 Sharpey 50 Torrington Square [London]	to Thomson

15 January 1878
Sharpey   to Thomson
University College [London]

15 January 1878
Sharpey   to Thomson
50 Torrington Square [London]

25 May 1878
Sharpey   to Thomson
50 Torrington Square [London]

17 July 1878
Sharpey   to Thomson
50 Torrington Sq [London]

12 September 1878
Sharpey   to Thomson
Wester Kinloch
Blairgowrie
Perthshire

16 February 1879
Sharpey   to Thomson
25 Wellington Sq
Hastings

10 March 1879
Sharpey   to Thomson
50 Torrington Square [London]

Wednesday
Sharpey   to Thomson
London

12 July [?]
Sharpey   to Thomson
33 Woburn Place
London

BOX 15: *Associated correspondence*

Letter 3:   [July 1836]
             Thomson   to Robert Carswell
             Campden Hill [London]

Letter 27:   6 February 1846 Friday afternoon
             Richard Quain   to Sharpey

Letter 26:   11 February 1846
             Sharpey   to James Syme
             London

Appendix

Letter 28:   11 February 1846
               Sharpey   to Lord Provost of Edinburgh
               London

Letter 31:   23 December 1846
               Richard Quain   to Sharpey

               30 March 1847
               William Hamilton   to Augustus de Morgan
               Edinburgh

               14 June 1849
               James Syme   to Allen Thomson
               Rutland St [Edinburgh]

               24 November 1853
               Arthur H. Hassall   to Sharpey
               8 Bennett St
               St. James St [London]
               [Copy of Sharpey's answer on back of this letter.]

               29 November 1853
               Arthur H. Hassall   to Sharpey
               8 Bennett St
               St. James St [London]

               13 December 1853
               Edward Sabine   to Sharpey
               Woolwich

               16 January 1854
               Edward Sabine   to Sharpey
               26 Medina [?]
               St. Leonards

               1 February 1854
               Arthur H. Hassall   to Sharpey
               8 Bennett St
               St. James St [London]

               4 February 1854
               Arthur H. Hassall   to Sharpey
               8 Bennett St
               St. James St [London]

               19 February 1854
               W. Bird Herapath   to Sharpey

               22 February 1854
               W. Bird Herapath   to Sharpey

               25 February 1854
               Dr Bird Herapath   to Sharpey

162

3 March 1854
Joseph Power   to Sharpey
Clare Hall
Cambridge

22 May 1854
C. Gyde   to Sharpey
Printing Office
Red Lion Court
Fleet St [London]

2 May 1859
Charles Babbage   to Benjamin Brodie
Dorset St
Manchester Sq [London]

12 August 1859
Benjamin C. Brodie   to Sharpey [?]

10 January 1860
Carl v. Siebold   to Sharpey
München

Letter 68:   16 January 1860
Thomson   to Benjamin Brodie
The College
Glasgow

15 January 1861
B[enjamin] C. Brodie   to Sharpey
14 Savile Row [London] W

15 January 1861
B[enjamin] C. Brodie   to Major-General [Edward] Sabine
14 Savile Row [London]

19 January 1861
B[enjamin] C. Brodie   to General [Edward] Sabine
14 Savile Row [London]

20 January 1861
Sharpey   to General [Edward] Sabine
33 Woburn Place
[London] W.C.

21 January 1861
B[enjamin] C. Brodie   to Major Gl. [Edward] Sabine
[London]

21 January 1861
B[enjamin] C. Brodie   to Sharpey
14 Savile Row [London] W

	21 June [?] 1861	
	B[enjamin] C. Brodie	to Sharpey
	– – – [?]	
	Betchworth [?]	
	Jersey [?]	
	30 December 1863	
	James Lehre [?]	to Thomson
	61 St. Vincent St	
	Glasgow	
	30 March 1864	
	John Burt	to Thomson
	Royal College of Physicians	
	Edinburgh	
	2 May 1864	
	Walker Arnott	to Thomson
	2 Victoria Terrace	
	Dowanhill	
	Glasgow	
Letter 79:	24 December 1866	
	Michael Foster	to Thomson
	Huntingdon	
	2 December 1867	
	John Storrar	to Thomson
	Heath Side	
	Hampstead	
	[London] NW	
	17 December 1867	
	John Storrar	to Thomson
	18 December 1867	
	Mary Colvill	to Thomson
	Lawnbank	
	Hampstead	
	[London] NW	
	21 December 1867	
	John Storrar	to Thomson
	24 December 1867	
	John Storrar	to Thomson
	31 December 1867	
	John Storrar	to Thomson
	9 July 1868	
	Mary Colvill	to Thomson
	Lawnbank [London]	

Appendix

21 August 1868
Henry B. Wheatley   to Thomson
The Royal Society
Burlington House
London W

24 August 1868
Mrs John Storrar   to Thomson
Ilkley

1873
[Draft letter from Sharpey to Edmondston and Douglas, publishers of a
life of J. Y. Simpson.]

1 March 1873
Mary Kelbe[?]   to Sharpey
1 Leigham Villas
Plymouth

8 May 1873
William Longman   to Sharpey
Paternoster Row [London]

Saturday 4 pm [10 May 1873]
John Marshall   to Thomson[?]
10 Savile Row [London]

29 May 1873
Frances Elizabeth Wade   to Sharpey
13 Richmond Terrace
Folkestone

16 June 1873
Henry Square   to Sharpey
West Terrace
Folkstone

12 July 1873
Sharpey   to Messrs Edmondston
                                        and Douglas

15 July 1873
Edmondston & Douglas   to Sharpey
88 Princes St   University College
Edinburgh   London

[8 November 1873]
Richard Quain   to Thomson
67 Harley St. [London] W

. *Letter 91*:   11 November 1873
                     Thomson   to Richard Quain
                     University Glasgow
                     [There are two letters to Quain with this date. One (not reproduced) is
                       marked *private*.]

165

12 November [1873]
Richard Quain   to Thomson
67 Harley St. [London] W

Letter 90:   November 1873
John Storrar   to Thomson

17 February [1874]
Richard Quain   to Thomson
67 Harley St [London] W

[September 1875]
E. A. Schafer   to Thomson
University College
[Date in Thomson's hand.]

13 September 1875
E. A. Schafer   to Sharpey
University College

15 September 1875
Sharpey   to E. A. Schafer
Grantown

16 February 1876
M. H. Allchin   to Sharpey
94 Wimpole St
[London] W

24 February 1876
E. A. Schafer   to Thomson
University College [London]

Letter 96:   19 August 1876
William Jenner   to Sharpey

20 October 1876
E. A. Schafer   to Thomson
University College [London]

2 November 1876
Ninian H. Thomson   to Thomson
117 Gloucester Terrace
Hyde Park
London

6 November 1876
E. A. Schafer   to Thomson
London

24 January 1877
E. A. Schafer   to Thomson
University College [London]

Appendix

8 April 1878
A. T. Scott   to Thomson
4 Anglesey Rd
St Leonards-on-Sea

7 September 1878
Walter T. Clegg   to Sharpey
64 Edge Lane
Liverpool

28 November 1878
John Storrar   to Thomson
24 Elgin Crescent
Kensington Park [London] W

10 February 1879
William H. Colvill   to Thomson
25 Wellington Sq
Hastings

[December 1879?]
John Burdon Sanderson   to Thomson
26 Gordon Square [London]

19 December 1879
J. Burdon Sanderson   to Thomson
26 Gordon Sq [London] W.C.

13 April 1880
T. P. Healey   to Thomson
63 Gloucester St
Gloucester Sq [London] W

17 April 1880
Albrecht von Koelliker   to Thomson
Dresden

1 May 1880
A. T. Scott   to Thomson

15 May 1880
Thomas Wharton-Jones   to Thomson
35 George St
Hanover Sq [London]

13 June 1880
John Williams   to Thomson
28 Harley St [London]

16 June 1880
Charles Wilson   to Thomson
7 Dalrymple Crescent
Edinburgh

1 September 1880
William Marshall   to Thomson
Barnes
London SW

2 September 1880
John G. Macvicar   to Thomson
The Manse of Moffat

27 September 1880
William Marshall   to Thomson
Barnes

2 October 1880
Robert Patterson   to Sharpey
32 Charlotte St
Leith

26 November 1880
William H. Colvill   to Thomson
9 Alexandra Place
Arbroath

6 December 1880
P. G. Tait   to Thomson
Royal Society of Edinburgh

11 December 1880
P. G. Tait   to Thomson
Royal Society of Edinburgh

22 December 1880
William H. Colvill   to Thomson
Alexandra Place
Arbroath

17 January 1881
George Burrows   to Thomson
18 Cavendish Sq [London]

22 January 1881
James Paget   to Thomson
Hanover Sq [London] W

3 February 1881
William H. Colvill   to Thomson
9 Alexandra Place
Arbroath

14 January 1881
Mrs J. Martin   to Thomson
Broughty Ferry   66 Palace Gardens Terrace
                                        London W

16 January 1881
George Harley   to Thomson
25 Harley St
Cavendish Sq [London] W

16 January 1881
William Aitken   to Thomson

16 January 1881
J. Matthews Duncan   to Thomson
71 Brook St   66 Palace Gdns
Grosvenor Sq [London] W Gardens Terrace
           Kensington W [London]

17 January 1881
R. Hassall   to Thomson
20 Lansdowne Road
Notting Hill [London]

18 January 1881
W. H. Colvill   to Thomson
Alexandra Place
Arbroath

18 January 1881
George Goodall   to Thomson
Pine Ridge
nr. Farnham
Surrey

22 January 1881
Mary Noble Salkirk   to Thomson
The Vicarage
Winchcombe
Gloucestershire

26 January 1881
John Williams   to Thomson
28 Harley St [London]

30 January 1881
W. Sharpey Seaton   to Thomson

11 October
Edward Sabine   to Dr Miller
Ashley Place

12 May
Edward Sabine   to Sharpey
The Athenaeum

[No date]
[Lord] Belper   to Sharpey
University College London
Gower St W.C.

[No date]
M. Thomson jun.   to Sharpey
Marland Cottage   2 Alva St

Thursday afternoon
E. A. Schafer   to Thomson
University College [London]

BOX 16: *Sharpey-Thomson letters*

16 May 1879
Thomson   to Sharpey
Park Villa
Woolston
Southampton

16 June 1879
Sharpey   to Thomson
50 Torrington Square [London]

14 July 1879
Sharpey   to Thomson
Wester Kinloch
Blairgowrie

11 August [1879]
Sharpey   to Thomson
Wester Kinloch
Blairgowrie

1 December 1879
Sharpey   to Thomson
31 Wellington Square
Hastings

8 December 1879
Sharpey   to Thomson
31 Wellington Square
Hastings

15 December 1879
Sharpey   to Thomson
31 Wellington Square
Hastings

16 January 1880
Sharpey   to Thomson
31 Wellington Square
Hastings

8 March 1880
Sharpey   to Thomson
31 Wellington Square
Hastings

14 March 1880
Thomson   to Sharpey
Florence

26 March 1880
Thomson   to Sharpey
Paris

BOX 16: *Associated correspondence*

27 December 1853
James Booth   to Thomas Bell
22 Sussex Gardens
Hyde Park [London]

24 February 1854
Sharpey   to Henry Bence Jones

24 February 1854
Sharpey   to W. B. Herrapath
London

27 September 1857
Stephen Place   to [Edward Sabine?]
at H. Cheswright
Villa Maggini
Torquay

9 September [1858]
Edward Sabine   to Sharpey

18 April 1874
Sharpey   to The President & Council
                                 of University College

6 October 1879
J. Martin   to Thomson
Wester Kinloch

21 November 1879
J. Martin   to Thomson
Wester Kinloch

10 April 1880
J. Martin   to Thomson
Dundee

12 April 1880
Thomson   to J. Martin
66 Palace Gardens Terrace
London W
[This letter was not sent.]

12 April 1980
J. B. Sanderson   to Thomson
26 Gordon Square [London]

13 April 1880
Mary Noble Jackson   to Thomson
The Vicarage
Winchcombe
Gloucestershire

16 April 1880
John Storrar   to Thomson
58 Marine Parade
Worthing

13 April [1880]
E. M. Thomson   to Thomson
17 Alva St [Edinburgh]

21 April [1880]
John Topham   to Richard Quain

22 April 1880
Michael Foster   to Thomson
Shelford

1 May 1880
Michael Foster   to Thomson
Shelford

1 May 1880
Mary Noble Jackson   to Thomson
The Vicarage
Winchcombe
Gloucestershire

5 May 1880
J. Martin   to Thomson
Embden House
Broughty Ferry

11 May 1880
James Arrott   to Thomson
148 Nethergate
Dundee

20 May [1880]
J. Martin   to Thomson
Broughty Ferry

2 June 1880
J. Small   to Thomson
University of Edinburgh

8 June 1880
H. N. Moseley   to Thomson
University of London
Burlington Gardens, [London] W

11 June 1880
James Arrott   to Thomson
148 Nethergate
Dundee

12 June 1880
William Henry Colvill   to Thomson
Marine Line
Bombay

15 June 1880
J. Martin   to Thomson
Western Kinloch
Blairgowrie

22 June 1880
Michael Foster   to Thomson
Shelford
Cambridgeshire

6 July 1880
J. Martin   to Thomson
Wester Kinloch

24 July 1880
Richard Quain   to Thomson
52 Cavendish Square [London] W

28 July [1880]
Richard Quain   to Thomson
32 Cavendish Square [London] W

18 August 1880
Lucy Storrar   to Thomson
Upper Urquhart
Strathmiglo
Fife

5 September 1880
William Henry Colvill   to Thomson
Philip St
Carnoustie

3 October 1880
William Henry Colvill   to Thomson
Ednburgh

20 October 1879
Thomas Henry Richards[?]   to Sharpey

25 October 1880
Michael Foster   to Thomson
Shelford
Cambridgeshire

8 November 1880
J. Martin   to Thomson
Wester Kinloch

9 November 1880
Richard Quain   to Thomson

12 November 1880
J. Martin   to Thomson
Wester Kinloch

14 November 1880
William Henry Colvill   to Thomson
Carnoustie

13 December [1880]
J. Martin   to Thomson
Wester Kinloch

17 December [1880]
J. Martin   to Thomson
Wester Kinloch

Christmas 1880
Helen Richard[?]   to Thomson
Embden House
Broughty Ferry

BOX 18

5 June 1860
Sharpey   to Thomson
33 Woburn Place
London W.C.

Index